SpringerBriefs in Mathemati

Series Editors

Nicola Bellomo, Torino, Italy

Michele Benzi, Pisa, Italy

Palle Jorgensen, Iowa City, USA

Roderick Melnik, Waterloo, Canada

Otmar Scherzer, Linz, Austria

Benjamin Steinberg, New York, NY, USA

Lothar Reichel, Kent, USA

Yuri Tschinkel, New York, NY, USA

George Yin, Detroit, USA

Ping Zhang, Kalamazoo, MI, USA

SpringerBriefs present concise summaries of cutting-edge research and practical applications across a wide spectrum of fields. Featuring compact volumes of 50 to 125 pages, the series covers a range of content from professional to academic. Briefs are characterized by fast, global electronic dissemination, standard publishing contracts, standardized manuscript preparation and formatting guidelines, and expedited production schedules.

Typical topics might include:

- A timely report of state-of-the art techniques
- A bridge between new research results, as published in journal articles, and a contextual literature review
- A snapshot of a hot or emerging topic
- An in-depth case study
- A presentation of core concepts that students must understand in order to make independent contributions

SpringerBriefs in Mathematics showcases expositions in all areas of mathematics and applied mathematics. Manuscripts presenting new results or a single new result in a classical field, new field, or an emerging topic, applications, or bridges between new results and already published works, are encouraged. The series is intended for mathematicians and applied mathematicians. All works are peer-reviewed to meet the highest standards of scientific literature.

Titles from this series are indexed by Scopus, Web of Science, Mathematical Reviews, and zbMATH.

Marian Mrozek • Thomas Wanner

Connection Matrices in Combinatorial Topological Dynamics

Springer

Marian Mrozek
Faculty of Mathematics and Computer Science, Department of Computational Mathematics
Jagiellonian University
Kraków, Poland

Thomas Wanner
Department of Mathematical Sciences
George Mason University
Fairfax, VA, USA

ISSN 2191-8198 ISSN 2191-8201 (electronic)
SpringerBriefs in Mathematics
ISBN 978-3-031-87599-1 ISBN 978-3-031-87600-4 (eBook)
https://doi.org/10.1007/978-3-031-87600-4

This Springer imprint is published by the registered company Springer Nature Switzerland AG
The registered company address is: Gewerbestrasse 11, 6330 Cham, Switzerland

To my beloved wife Joanna, children Maria,
Krzysztof, Marek, Joanna, and grandchildren
Amelia, Sara, Magdalena, Weronika.
And in memory of my parents Maria and Jan

Marian Mrozek

To Evelyn, Miriam, and Alice

Thomas Wanner

Preface

The purpose of this brief is to introduce the theory of connection matrices in the context of combinatorial dynamical systems. Connection matrices are one of the central tools in Conley's approach to the study of classical dynamical systems, as they provide information on the existence of connecting orbits in Morse decompositions. They may be considered as a generalization of the Morse complex boundary operator in Morse theory. Unfortunately, however, the theory of connection matrices is scattered throughout numerous research papers. This makes it difficult for the novice to learn or apply these techniques, and to uncover new areas and directions for their use.

Recently, some aspects of Conley's dynamical theory have been carried over to the combinatorial setting, in the context of multivector fields on Lefschetz complexes. This brief grew out of our attempt to also adapt the theory of connection matrices to this combinatorial context. While parts of their construction can be developed fairly easily from classical results, other parts required significant new ideas such as connection matrices based on varying posets, as well as formally defining the concept of multiple connection matrices. These later concepts lead to new insights even in the classical situation. Nevertheless, the overall combinatorial setting allows for a more streamlined presentation of the theory, which avoids some of the technical difficulties of previous approaches. In view of this, the present book provides a complete and self-contained introduction to the subject. It is our hope that this will make it easier to learn about this fascinating and deep topic, and that it will lead to novel application areas.

Kraków, Poland — Marian Mrozek
Fairfax, VA, USA — Thomas Wanner
February 2025

Acknowledgments

Throughout the preparation of this work, we have benefitted from numerous conversations with colleagues and students. We would like to specifically acknowledge our discussions with Tamal Dey, Herbert Edelsbrunner, Bartosz Furmanek, Shaun Harker, Michał Lipiński, Filip Łanecki, Konstantin Mischaikow, Kelly Spendlove, and Justin Thorpe.

In addition, the research of Marian Mrozek was partially supported by the Polish National Science Center under Maestro Grant No. 2014/14/A/ST1/00453 and Opus Grant No. 2019/35/B/ST1/00874. Thomas Wanner was partially supported by NSF grant DMS-1407087 and by the Simons Foundation under Award 581334.

Competing Interests The authors have no competing interests to declare that are relevant to the content of this manuscript.

Contents

Chapter 1
Introduction

In this chapter, we briefly provide context for classical Morse and Conley theory and describe in particular the standard pipeline for computing connection matrices. We then discuss the motivation for combinatorial topological dynamics based on multivectors and outline the overall structure of the present book.

1.1 Connection Matrices in the Classical Setting

Classical Morse theory concerns a compact smooth manifold together with a smooth real-valued function with non-degenerate critical points. An important part of the theory introduces the Morse complex which is a chain complex whose k-th chain group is a free abelian group spanned by critical points of Morse index k and whose boundary homomorphism is defined by counting the (oriented) flow lines between critical points in the gradient flow induced by the Morse function (see [26, Section 4.2]). One of the fundamental results of classical Morse theory states that the homology of the manifold is isomorphic to the homology of the Morse complex.

The stationary points of the gradient flow in Morse theory, which are precisely the critical points of the Morse function, provide the simplest example of an isolated invariant set, a key concept of Conley theory [11]. For every isolated invariant set, there is a homology module associated with it. It is called the homological Conley index. A (minimal) Morse decomposition is a decomposition of space into a partially ordered collection of isolated invariant sets, called Morse sets, such that every recurrent trajectory (in particular every stationary or periodic trajectory) is located in a Morse set and every nonrecurrent trajectory is a heteroclinic connection between Morse sets from a higher Morse set to a lower Morse set in the poset structure of the Morse decomposition. The collection of stationary points of the gradient flow of a Morse function provides the simplest example of a Morse decomposition in which the Morse sets are just the stationary points and the Conley

M. Mrozek, T. Wanner, *Connection Matrices in Combinatorial Topological Dynamics*, SpringerBriefs in Mathematics,
https://doi.org/10.1007/978-3-031-87600-4_1

index of a stationary point coincides with the homology of a pointed k-dimensional sphere with k equal to the Morse index of the point.

Conley theory in its simplest form may be viewed as a twofold generalization of Morse theory. On the one hand, it substantially weakens the general assumptions through replacing the smooth manifold by a compact metric space and the gradient flow of the Morse function by an arbitrary (semi)flow. On the other hand, it replaces the collection of critical points of the Morse function by the more general Morse decomposition in which the counterpart of the Morse complex takes the form of the direct sum of the Conley indices of all Morse sets. The homology of this generalized complex, as in Morse theory, is isomorphic to the homology of the underlying space. The boundary operator in this setting is called the connection matrix of the Morse decomposition, and it was introduced by Franzosa in [20]. His definition, based on homology braids, is technically complicated, in part because the generalized setting captures the situations of bifurcations when, unlike for Morse theory, the connection matrix need not be uniquely determined by the flow. In a later paper, Robbin and Salamon [44] slightly simplify the definition by replacing homology braids with filtered chain complexes which helps with separating dynamics from algebra. This separation is even more visible in the recent algorithmic approach to connection matrices by Harker et al. [21, 46]. In fact, the separation of dynamics and algebra allows the authors in [21, 44, 46] to set up the definition of a connection matrix of a Morse decomposition along the following pipeline, where we omit a number of technical details:

(i) Consider a Morse decomposition $\mathcal{M} := \{M_p\}_{p\in P}$ of the phase space which is indexed by a poset P.
(ii) For each down set $I \subset P$, consider the associated attractor M_I consisting of points on trajectories whose limits sets are in $\bigcup_{p\in I} M_p$, and construct an attracting neighborhood N_I for the attractor M_I in such a way that $I \mapsto N_I$ is a lattice homomorphism.
(iii) Under some smoothness assumptions, the attracting neighborhood family $\{N_I\}$ of step (ii) then induces a P-filtered chain complex.
(iv) A connection matrix of $\mathcal{M}$ is an algebraic object associated with a poset filtered chain complex, in particular with the P-filtered chain complex of step (iii).

It will become clear later on that the results presented in this brief enable us to shorten this pipeline by essentially eliminating step (ii). This will be described in some detail in the next section, as well as in Chap. 2.

1.2 Combinatorial Topological Dynamics

Conley theory, in particular via connection matrices, is a very useful tool in the qualitative study of dynamical systems. However, to apply it, one requires a well-defined dynamical system on a compact metric space. This is not the case when the dynamical system is exposed only via a finite set of samples as in the case of time series collected from observations or experiments. The study of dynamical systems known only from samples becomes an important part of the rapidly growing field of data science. In this context, a generalization of Morse theory presented by Robin Forman [19] turns out to be very fruitful. In his generalization, the smooth manifold is replaced by a finite CW complex and the gradient vector field of the Morse function by the concept of a combinatorial vector field. These structures may be easily constructed from data and analyzed by means of the combinatorial, also called discrete, Morse theory by Forman.

Recently, the concepts of isolated invariant set and Conley index have been carried over to this combinatorial setting [5, 25, 30, 35, 38]. In the present book, we extend these ideas by constructing connection matrices of a Morse decomposition of a combinatorial multivector field. To achieve this, one has to modify the connection matrix pipeline discussed above, because in the combinatorial case, and even in the case of a multiflow, step (ii) in the pipeline cannot be completed in general. To overcome this difficulty, we enlarge the original poset to guarantee the existence of the necessary lattice of attracting neighborhoods. The added elements are then removed by introducing a certain equivalence in the category of poset filtered chain complexes with a varying poset structure. A side benefit of this approach is the shortening of the pipeline by omitting step (ii). Such a shortening is possible, because, under the enlarged poset, the filtered chain complex of step (iii) may be obtained directly from the Morse decomposition via an associated partition of the phase space indexed by the extended poset.

Although the main motivation for our book is the adaptation of connection matrix theory to the combinatorial setting of multivector fields, we believe that the approach presented here has broader potential. Formally speaking, connection matrices are purely algebraic objects and are presented this way from the very beginning. Nevertheless, in the early papers, they are strongly tied to dynamical considerations. In particular, the important concept of uniqueness and nonuniqueness in these papers is addressed only via the underlying dynamics. As we already mentioned, the decoupling of algebra and dynamics started with Robbin and Salamon [44] and is even stronger in Harker et al. [21]. However, to the best of our knowledge, so far there has been no purely algebraic definition of uniqueness. Harker et al. [21] prove that any two connection matrices of the same filtered chain complex are conjugate via a filtered isomorphism. Clearly, this is not the uniqueness concept used in the context of dynamics. In our approach, we propose a stronger algebraic equivalence of connection matrices which allows for filtered chain complexes with no unique connection matrix. In this way, the proposed theory of connection matrices for combinatorial multivector fields naturally allows us to further prove an important new result of the book: The uniqueness of connection matrices for gradient combinatorial Forman vector fields, which is based solely on the above-mentioned stronger notion of equivalence.

We believe that the detachment of connection matrix theory from dynamics is worth the effort, because it may bring applications in new fields. The potential of applications in topological data analysis is already discussed in [21], and the ties between connection matrices and persistent homology are also visible in [13].

Moreover, combinatorial vector fields have been used successfully in the study of both algebraic and combinatorial problems [22, 27, 45].

Finally, combinatorial multivector fields make it possible to construct examples of a variety of complex dynamical phenomena in a straightforward way. It is therefore our hope that the results of this book make topological methods in dynamics, and in particular the concept of connection matrices, more accessible to a broader mathematical audience.

1.3 Organization of the Book

The remainder of this book is organized as follows. In Chap. 2, we give an informal overview over our combinatorial theory of connection matrices. By providing numerous examples, we hope that this chapter will serve as a guide for the more technical later chapters of the book. In addition, this chapter can serve as a quick and intuitive introduction to the concept of connection matrices. After presenting necessary preliminary definitions and background results from both algebra and topology in Chap. 3, the main technical part of the book starts with a discussion of poset graded and poset filtered chain complexes in Chap. 4. Based on these definitions, the algebraic connection matrix can then be introduced in Chap. 5. The remaining three chapters of the book apply the earlier algebraic constructions to multivector fields on Lefschetz complexes. While the connection matrix in this setting is considered in Chap. 6, the dynamics of combinatorial multivector fields is the subject of Chap. 7. Finally, in Chap. 8, we show that if one considers a gradient Forman vector field on a Lefschetz complex, then the associated connection matrix is necessarily uniquely determined. This last chapter closes with a brief discussion of open problems and potential directions for future extensions of the theory.

Chapter 2
A Guided Tour of Connection Matrices

In this chapter, we provide a guided tour through our theory of combinatorial connection matrices, supplemented by numerous examples. This includes a summary of our definition of connection matrices, as well as outlines of the main results of the book. We also introduce the necessary background material, albeit not in full technical detail. In this way, the current chapter serves as a quick introduction to the theory, while avoiding some of the more technical aspects. Precise definitions, statements, and proofs will be given in the sequel. In addition, we briefly address the computability of connection matrices and demonstrate how in combination with the theory of multivector fields they can be used to analyze the global dynamics of low-dimensional ordinary differential equations.

2.1 Combinatorial Dynamical Systems and Multivector Fields

By a *combinatorial dynamical system*, we mean a multivalued map $F : X \multimap X$ defined on a finite topological space X. Alternatively, it may be viewed as a finite directed graph whose set of vertices is the topological space X, and with F interpreted as the map sending a vertex to the collection of its neighbors connected via an outgoing directed edge. We call it the *F-digraph*. In most applications, this set X of vertices of the F-digraph is a collection or certain subcollection of cells of a finite cellular complex, for instance a simplicial complex, with its topology induced by the associated face poset via the Alexandrov theorem [1]. This topology is very different from the metric topology of the geometric realization of the cellular complex in terms of its separation properties, but the same in terms of homotopy and homology groups via McCord's theorem [33]. Hence, as far as algebraic topological invariants are concerned, these topologies may be used interchangeably, with the

M. Mrozek, T. Wanner, *Connection Matrices in Combinatorial Topological Dynamics*, SpringerBriefs in Mathematics,
https://doi.org/10.1007/978-3-031-87600-4_2

finite topology having the advantage of better explaining some peculiarities of combinatorial dynamics.

As in classical multivalued dynamics, given a combinatorial dynamical system F, we define a *solution* of F to be a map $\gamma : I \to X$ defined on a subset $I \subset \mathbb{Z}$ of integers and such that $\gamma(i+1) \in F(\gamma(i))$ for $i, i+1 \in I$. The solution γ is called *full* if $I = \mathbb{Z}$, and it is a *path* if I is the intersection of $\mathbb{Z}$ with a compact interval in $\mathbb{R}$. In the directed graph interpretation, a solution may be viewed as a finite or infinite directed walk through the graph. Finally, a subset $A \subset X$ is called *invariant* if for every point $a \in A$ there exists a full solution $\gamma : \mathbb{Z} \to A$ through the point a, that is, a solution satisfying the identity $\gamma(0) = a$.

In this book, we are interested in a class of combinatorial dynamical systems induced by combinatorial vector and multivector fields. We recall that a *combinatorial multivector field* $\mathcal{V}$ is a partition of a finite topological space X into multivectors, where a *multivector* is a *locally closed set*, that is, a set that is the difference of two closed sets (see Proposition and Definition 3.2.1). A multivector is called a *vector*, if its cardinality is one or two. If a combinatorial multivector field contains only vectors, then we call it a *combinatorial vector field*, a concept introduced already by Forman [19]. In applications, the partition indicates the resolution below which there is not enough evidence to analyze the dynamics either due to an insufficient amount of data or because of lacking computational power.

The results of this book require the combinatorial multivector field to be defined on a special finite topological space, namely a Lefschetz complex [29], called by Lefschetz and in [21] simply a complex. In short, a *Lefschetz complex* is just a basis X of a finitely generated free chain complex (C, d) with emphasis on the basis. This means that X is the Lefschetz complex, and (C, d), denoted $C(X)$, is the chain complex associated with the Lefschetz complex. In terms of applications, a typical example of a Lefschetz complex is the set of cells of a cellular complex or simplices of a simplicial complex. These simplices or cells constitute a natural basis for the associated chain complex. Also, every locally closed subset of a Lefschetz complex is a Lefschetz complex, a feature that, in particular, facilitates constructing concise examples. By the homology of a Lefschetz complex X, we mean the homology of the associated chain complex $C(X)$. We view a Lefschetz complex as a finite topological space via the Alexandrov theorem. For this, one needs a well-defined face relation, which is given by the transitive closure of the facet relation, where an element $x \in X$ is called a *facet* of $y \in X$ if x appears with nonzero coefficient in the boundary $d(y)$.

Example 2.1.1 (A First Multivector Field, $\triangleright$ 2.1.2 [1]**)** The left image in Fig. 2.1 presents an example of a combinatorial multivector field on a Lefschetz complex X which is just a simplicial complex consisting of the two triangles **ABC**, **BCD**, the five edges **AB**, **AC**, **BC**, **BD**, **CD**, and the four vertices **A**, **B**, **C**, **D**. Hence, X is a finite topological space consisting of eleven elements. The combinatorial

[1] Here and in the remainder of the book, the sign $\triangleright$ should be read as *continued as Example*, and the sign $\triangleleft$ should be read as *continued from Example*.

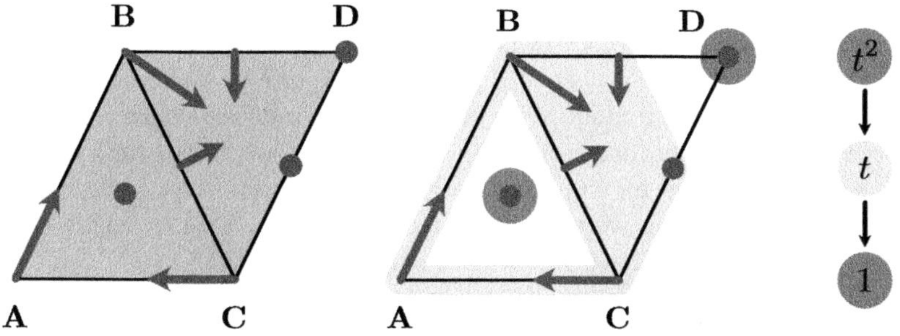

Fig. 2.1 *A first multivector field.* The left panel shows a combinatorial multivector field on a simplicial complex that consists of two triangles, five edges, and four vertices. The multivector field has three critical cells, marked as red dots, two vectors on the bottom and left-most edges, as well as one multivector, which consists of the right triangle **BCD**, its two edges **BC** and **BD**, and the vertex **B**. In the middle panel, we indicate the three Morse sets of the associated combinatorial dynamical system $F_{\mathcal{V}}$ in purple, yellow, and green, while the Conley-Morse graph of this Morse decomposition is shown in the panel on the right

multivector field shown in Fig. 2.1 consists of the three singletons {**ABC**}, {**CD**}, and {**D**} indicated in the figure by a red dot, the two doubletons {**A**, **AB**} and {**C**, **AC**} marked with a red arrow, and one multivector {**B**, **BC**, **BD**, **BCD**} consisting of four simplices and marked with red arrows joining each simplex in the multivector with its top-dimensional coface in the same multivector. Why this method of drawing multivectors is natural will become clear when we discuss the combinatorial dynamical system induced by the combinatorial multivector field. ◊

Every combinatorial multivector field $\mathcal{V}$ induces a combinatorial dynamical system $F_{\mathcal{V}} : X \multimap X$ given for $x \in X$ by the multivalued map

$$F_{\mathcal{V}}(x) := \operatorname{cl} x \cup [x]_{\mathcal{V}} . \tag{2.1}$$

In this definition, $\operatorname{cl} x$ stands for the closure of x in X, that is, it is given by the collection of all faces of x. Moreover, $[x]_{\mathcal{V}}$ denotes the unique multivector V in the partition $\mathcal{V}$ with $x \in V$. The formula for $F_{\mathcal{V}}$ is related to the resolution interpretation of a combinatorial multivector field mentioned earlier. Namely, inside a multivector, we do not exclude any movement, treating a multivector as kind of a black box. This justifies the presence of $[x]_{\mathcal{V}}$ in (2.1). And, since the dynamics of a combinatorial multivector field models a flow, any movement between multivectors is possible only through their common boundary. This justifies the presence of the closure $\operatorname{cl} x$ in the definition (2.1).

Example 2.1.2 (A First Multivector Field, ◁ 2.1.1 ▷ 2.1.3) Even for a simple multivector field such as the one shown in the left panel of Fig. 2.1, the induced combinatorial dynamical system is quite large. Interpreted as a directed graph, it has 11 vertices (one for each simplex) and 42 directed edges. Hence, instead of drawing such a digraph, we interpret Fig. 2.1 as a digraph with vertices in the centers of mass of simplices (not marked) and only a minimum of directed edges (arrows) which cannot be deduced from (2.1). In fact, the only arrows which do depend on $\mathcal{V}$ are arrows of the form $x \to y$ where $y \in [x]_{\mathcal{V}} \setminus \operatorname{cl} x$. Moreover, since vertices inside the same multivector form a clique in the directed graph, it suffices to mark only some of the arrows joining them, and, as we already mentioned, we mark arrows joining each simplex in the multivector with its top-dimensional cofaces in the same multivector. Interpreting Fig. 2.1 this way, it is not difficult to infer the solutions of $F_{\mathcal{V}}$ from the figure. For instance, we have a solution γ defined for $n \in \mathbb{Z}$ by

$$\gamma(n) := \begin{cases} \mathbf{ABC} & \text{for } n < 0, \\ \mathbf{BC} & \text{for } n = 0, \\ \mathbf{BCD} & \text{for } n = 1, \\ \mathbf{BD} & \text{for } n = 2, \\ \mathbf{D} & \text{for } n > 2. \end{cases}$$

In the sequel, we will write solutions in a compact form

$$\gamma = \overleftarrow{\mathbf{ABC}} \cdot \mathbf{BC} \cdot \mathbf{BCD} \cdot \mathbf{BD} \cdot \overrightarrow{\mathbf{D}}$$

where the left and right arrows on top of the left and right ends, respectively, indicate that the same pattern is repeated up to negative and positive infinity. ◊

An unintended consequence of the otherwise natural formula (2.1) is that regardless of the specific choice of multivector field $\mathcal{V}$, every point $x \in X$ is a fixed point of $F_{\mathcal{V}}$, that is, we have $x \in F_{\mathcal{V}}(x)$. This immediately implies that every subset of X is invariant for combinatorial dynamical systems induced by combinatorial multivector fields. Clearly, the interest is in invariant sets whose invariance goes beyond the fact that all points are fixed points of $F_{\mathcal{V}}$. In order to properly describe such invariant sets, we need the following observations and definitions.

According to (2.1), a solution of the combinatorial dynamical system may leave a multivector $V \in \mathcal{V}$ only via a point in $\operatorname{mo} V := \operatorname{cl} V \setminus V$, a set which we call the *mouth* of V. Indeed, if $\sigma(i) \in V$ and $\sigma(i+1) \notin V$, then necessarily $\sigma(i+1) \in \operatorname{mo} V$. This means that the mouth of V acts as the *exit set* for V. The mouth is closed, because V is locally closed. In classical Conley theory, a closed set N whose exit set N^- is closed as well is called an isolating block, and the relative homology $H(N, N^-)$ is the associated *Conley index*, see, for example, [11, 47]. We adopt this terminology, and, interpreting the closure $\operatorname{cl} V$ as a small *isolating block* for the multivector V, we call the relative homology $H(\operatorname{cl} V, \operatorname{mo} V)$ the *Conley index* of V. In the classical setting, a nontrivial Conley index guarantees the existence of a solution which stays within the isolating block V for all times, while a trivial index allows for the flow to merely pass through V. Based on this, we call a multivector *critical* if we have the inequality $H(\operatorname{cl} V, \operatorname{mo} V) \neq 0$, and *regular* otherwise. Motivated by the fact that in the classical setting a zero Conley index of an isolating block does not guarantee the existence of a full solution inside the block, we say that a full solution γ is *essential*, if for every regular multivector V the preimage $\gamma^{-1}(V)$ does not contain an infinite interval of integers. In other words, every essential solution must leave a regular multivector V in forward and backward time before it can reenter the multivector.

An invariant set A is called an *essential invariant set*, if every $a \in A$ admits an essential solution through a which is completely contained in A. We are interested in *isolated invariant sets*, which are defined as essential invariant sets admitting a closed superset $N \supset S$ such that $F(S) \subset N$, and such that every path in N with endpoints in S is itself contained in S. We call such a set N an *isolating set*. Isolating sets are the combinatorial counterparts of isolating neighborhoods in the classical setting. Nevertheless, we prefer the name *isolating set*, because, in the combinatorial setting, they do not need to be neighborhoods in general. An isolating set is called an *attracting set* if every path in X with left endpoint in N is contained in N, and the associated isolated invariant set S is then called an *attractor*. Similarly, one defines a *repelling set* and a *repeller*, by replacing the left endpoint in the above definitions with the right endpoint. We note that an essential invariant set is an isolated invariant set if and only if it is locally closed and *$\mathcal{V}$-compatible*, that is, it equals the union of all multivectors contained in it, see, for example, [30, Proposition 4.10, 4.12, 4.13]. Also, an essential invariant set is an attractor (or repeller) if and only if it is $\mathcal{V}$-compatible and closed (or open), as was shown in [30, Theorem 6.2].

Example 2.1.3 (A First Multivector Field, ◁ 2.1.2 ▷ 2.2.1) It is easy to verify that every singleton in a multivector field is critical, and every doubleton is necessarily regular. In contrast, for a general multivector of cardinality greater than two, it is not possible to automatically determine whether it is critical or regular just based on its cardinality. For example, in Fig. 2.1, the multivector $V := \{\mathbf{B}, \mathbf{BC}, \mathbf{BD}, \mathbf{BCD}\}$ is regular, since its closure may be homotopied to its mouth, that is, one has $H(\operatorname{cl} V, \operatorname{mo} V) = 0$. Yet, it is not difficult to see that detaching the vertex $\mathbf{B}$ from V and instead attaching the edge $\mathbf{CD}$ to V preserve the cardinality of the multivector but make V critical. An example of a solution in Fig. 2.1 which is not essential is

$$\overleftarrow{\mathbf{C} \cdot \mathbf{AC}} \cdot \mathbf{A} \cdot \mathbf{AB} \cdot \overrightarrow{\mathbf{B} \cdot \mathbf{BCD}}$$

and of a solution which is essential is

$$\overleftarrow{\mathbf{C} \cdot \mathbf{AC} \cdot \mathbf{A} \cdot \mathbf{AB} \cdot \mathbf{B} \cdot \mathbf{BCD}} \cdot \overrightarrow{\mathbf{D}}\,.$$

Furthermore, if we consider $A := \{\mathbf{A}, \mathbf{B}, \mathbf{C}, \mathbf{AB}, \mathbf{AC}, \mathbf{BC}\}$ in Fig. 2.1, then the solution

$$\overleftrightarrow{\mathbf{A} \cdot \mathbf{AB} \cdot \mathbf{B} \cdot \mathbf{BC} \cdot \mathbf{C} \cdot \mathbf{AC}} \tag{2.2}$$

is a periodic essential solution which passes through all elements of A. This demonstrates that the set A is an essential invariant set. However, it is not an isolated invariant set, because it is not $\mathcal{V}$-compatible. We may modify the set A by considering instead $A' := A \cup \{\mathbf{BD}, \mathbf{BCD}\}$. Then the enlarged set A' is $\mathcal{V}$-compatible, but still not an isolated invariant set, since it is not locally closed. In fact, the smallest isolated invariant set containing A is given by $A'' := A \cup \{\mathbf{BD}, \mathbf{BCD}, \mathbf{CD}\}$, which contains all simplices of X except for the triangle **ABC** and the vertex **D**. ◊

2.2 Conley Index and Morse Decompositions

Given an isolated invariant set S of a combinatorial multivector field $\mathcal{V}$ on a Lefschetz complex X, we define its *Conley index* as the relative homology $H(\operatorname{cl} S, \operatorname{mo} S)$. This definition is motivated by the fact that the topological pair $(\operatorname{cl} S, \operatorname{mo} S)$ is one of possibly many index pairs for S, see, for example, [30, Definition 5.1 and Proposition 5.3] for more details. Since in this chapter we are only interested in homology with field coefficients, and since such homology is uniquely determined by the associated Betti numbers, it is convenient to identify the homology and, in particular, the Conley index with the associated *Poincaré polynomial* whose coefficients are the consecutive Betti numbers. In the case of the Conley index, we refer to this polynomial as the *Conley polynomial* of the isolated invariant set.

Example 2.2.1 (A First Multivector Field, ◁ 2.1.3 ▷ 2.2.2) In the situation of the combinatorial multivector field depicted in Fig. 2.1, the multivector given by the singleton set $S := \{\mathbf{ABC}\}$ is an isolated invariant set. One can easily see that the relative homology $H(\operatorname{cl} S, \operatorname{mo} S)$ is the homology of a pointed 2-sphere. Therefore, the Conley polynomial of $\{\mathbf{ABC}\}$ is given by t^2. Similarly, it was shown in Example 2.1.3 that the smallest isolated invariant set containing the periodic solution (2.2) is given by the set A'' of all simplices of X, except for the triangle **ABC** and the vertex **D**. We leave it for the reader to verify that the Conley index $H(\operatorname{cl} A'', \operatorname{mo} A'')$ then has the Conley polynomial t. Finally, the Conley index of the singleton $\{\mathbf{D}\}$ is the homology of a pointed 0-sphere, and therefore the associated Conley polynomial is 1. ◊

An indexed family $(M_r)_{r \in P}$ of mutually disjoint isolated invariant sets indexed by a poset P is called a *Morse decomposition* of X if for every essential solution γ either all values of γ are contained in the same set M_r, or there exist indices $q > p$ in P and $t_p, t_q \in \mathbb{Z}$ such that $\gamma(t) \in M_q$ for $t \le t_q$ and $\gamma(t) \in M_p$ for $t \ge t_p$. In the latter case, the solution γ is called a *connection* from M_q to M_p. Furthermore, the sets M_r are called the *Morse sets* of the Morse decomposition.

One can show that the collection of all strongly connected components of the F-digraph which contain an essential solution does indeed form a Morse decomposition. In fact, in our setting, this Morse decomposition is always the finest one, see, for example, [12, Theorem 4.1]. In this respect, the combinatorial case differs from the classical situation where the finest Morse decomposition may not exist.

It is customary to condense the information about a Morse decomposition in the form of its *Conley-Morse graph*. This graph consists of the Hasse diagram of the poset P whose vertices, representing the individual Morse sets M_p, are labeled with their respective Conley polynomials.

Example 2.2.2 (A First Multivector Field, $\lhd$ 2.2.1 $\rhd$ 2.3.1) Consider $P := \{\mathbf{p}, \mathbf{q}, \mathbf{r}\}$ linearly ordered by $\mathbf{p} < \mathbf{q} < \mathbf{r}$, and again the combinatorial multivector field presented in the left image of Fig. 2.1. It is not difficult to verify that the family $\mathcal{M}$ consisting of $M_{\mathbf{r}} = \{\mathbf{ABC}\}$, $M_{\mathbf{q}} = \{\mathbf{A}, \mathbf{B}, \mathbf{C}, \mathbf{AB}, \mathbf{AC}, \mathbf{BC}, \mathbf{BD}, \mathbf{CD}, \mathbf{BCD}\}$, and $M_{\mathbf{p}} = \{\mathbf{D}\}$ is a Morse decomposition. In the middle panel of Fig. 2.1, the three Morse sets are indicated with three different colors. In our example all three Morse sets have one-dimensional Conley indices, respectively, in dimension zero for $M_{\mathbf{p}}$, one for $M_{\mathbf{q}}$, and two for $M_{\mathbf{r}}$. Hence, the Conley polynomials are given by 1 for $M_{\mathbf{p}}$, t for $M_{\mathbf{q}}$, and t^2 for $M_{\mathbf{r}}$. The associated Conley-Morse graph is visualized in the right-most panel of Fig. 2.1.

Although we have not yet given the definition of the connection matrix for a Morse decomposition, we can characterize some of its features on the basis of this example. As we will see later, a connection matrix is just a matrix representation of an abstract boundary homomorphism acting on the direct sum of Conley indices of Morse sets. The entries in the matrix are homomorphisms between the individual Conley indices. As in the classical Morse theory, the homology of the resulting chain complex coincides with the homology of the underlying Lefschetz complex. Assuming homology coefficients in the field $\mathbb{Z}_2$, the only possible homomorphisms between the one-dimensional Conley indices of Morse sets in our example are either zero or an isomorphism. Denote the Conley index of the Morse set M_p by C_p. Then the connection matrix for our example turns out to be

	$C_{\mathbf{p}}$	$C_{\mathbf{q}}$	$C_{\mathbf{r}}$
$C_{\mathbf{p}}$		0	
$C_{\mathbf{q}}$			1
$C_{\mathbf{r}}$			

,

where isomorphisms or the zero map is indicated by 1 or 0, respectively, and all empty fields automatically denote the zero homomorphism.[2] Hence, all entries of the connection matrix are zero, except for the homomorphism from $C_{\mathbf{r}}$ to $C_{\mathbf{q}}$. In our simple example, each Conley index is nontrivial in a different grade, and this in turn implies that $C_{\mathbf{p}}$, $C_{\mathbf{q}}$, and $C_{\mathbf{r}}$ are also chain subgroups in grades zero, one, and two,

[2] It will be useful at times to specifically highlight certain zero homomorphisms in a connection matrix, even though we could just have left the corresponding entry blank.

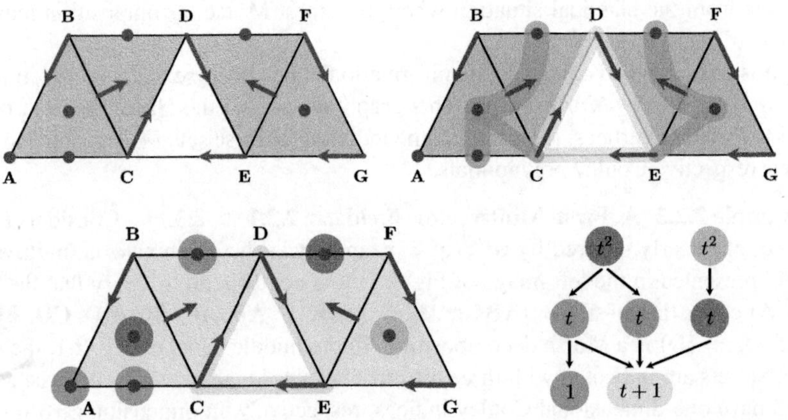

Fig. 2.2 *A Forman vector field with periodic orbit.* The top left panel depicts a Forman vector field on a simplicial complex, which consists of six critical cells and eight vectors given by doubletons. Notice that the edges **CD**, **DE**, and **CE** give rise to a periodic solution. In the top right panel, five connections originating in the two critical cells in dimension two are shown. While three of these connect to critical cells of dimension one, two connect to the periodic orbit. The two panels on the bottom illustrate the Morse sets of the combinatorial vector field in different colors, together with the associated Conley-Morse graph

respectively. Therefore, the chain complex may be written in the compact form

$$\cdots \xleftarrow{0} 0 \xleftarrow{0} C_{\mathbf{p}} \xleftarrow{0} C_{\mathbf{q}} \xleftarrow{1} C_{\mathbf{r}} \xleftarrow{0} 0 \xleftarrow{0} \cdots .$$

One can easily see that its homology is trivial except in grade zero—and this provides precisely the homology of the Lefschetz complex in the considered example. In fact, the connection matrix acts as an algebraic version of the Conley-Morse graph. Its unique nonzero entry reflects the heteroclinic connection from the Morse set $M_{\mathbf{r}}$ to the set $M_{\mathbf{q}}$. Notice that we also have two different heteroclinic connections from the Morse set $M_{\mathbf{q}}$ to the set $M_{\mathbf{p}}$, given, for example[3], by

$$\overleftarrow{\mathbf{CD}} \cdot \overrightarrow{\mathbf{D}} \qquad \text{and} \qquad \overleftarrow{\mathbf{CD}} \cdot \mathbf{C} \cdot \mathbf{AC} \cdot \mathbf{A} \cdot \mathbf{AB} \cdot \mathbf{B} \cdot \mathbf{BD} \cdot \overrightarrow{\mathbf{D}},$$

but they algebraically annihilate each other and thus lead to the corresponding entry being zero in the connection matrix. ◊

Example 2.2.3 (A Forman Vector Field with Periodic Orbit, ▷ 2.6.4) An example with a more elaborate Morse decomposition can be found in Fig. 2.2.

[3] Notice that since the dynamics within a multivector is not uniquely determined, there can be many different representations for the same solution. For example, the second of the above two heteroclinics could also be given as the solution $\overleftarrow{\mathbf{CD}} \cdot \mathbf{C} \cdot \mathbf{AC} \cdot \mathbf{A} \cdot \mathbf{AB} \cdot \mathbf{B} \cdot \mathbf{BC} \cdot \mathbf{BCD} \cdot \overrightarrow{\mathbf{D}}$. This solution traverses the same sequence of multivectors, but using different simplices.

In this case, the underlying Lefschetz complex is the simplicial complex indicated in the top left panel of the figure. The multivector field $\mathcal{V}$ is actually a Forman combinatorial vector field, which consists of six critical cells, together with eight vectors given by doubletons. Of the critical cells, two have Morse index 2, three have Morse index 1, while only one has index 0—where the Morse index of a critical cell is defined as the degree of its Conley polynomial. We would like to point out that in the case of a critical cell, this polynomial is always a monomial. In addition to the critical cells, the edges **CD**, **DE**, and **CE** give rise to a periodic solution. Together, these seven sets constitute the Morse sets of the minimal Morse decomposition of the vector field $\mathcal{V}$. They are indicated in different colors in the panel on the lower left. In order to get the full structure of the Morse decomposition, one needs to also determine the partial order between these Morse sets. For this, the top right panel shows five connections originating in the two critical cells in dimension two. While three of these connect to critical cells of dimension one, two connect to the periodic orbit. Similarly, one can easily determine which critical cells of Morse index 1 connect to the periodic orbit or the vertex **A**. From this, one can readily determine the Conley-Morse graph shown in the lower right panel. ◊

Intuitively, one can immediately see the connecting orbit structure of the above example, and therefore also its associated Morse decomposition. In fact, it was shown in [38] that for any Forman vector field on a simplicial complex X, one can explicitly construct a classical semiflow on X which exhibits the same Morse sets and Conley-Morse graph. Conversely, it was shown in [37] that suitable phase space subdivisions in a classical dynamical system combined with ideas from multivector fields can be used to rigorously establish the existence of classical periodic solutions from the existence of combinatorial counterparts. These results show that the above-mentioned intuition is more than just a coincidence.

Example 2.2.4 (A Multiflow Without Lattice of Attractors, ▷ 2.2.5) As our next example, consider the combinatorial multivector field presented in the left part of Fig. 2.3. It is defined on the Lefschetz complex X which is obtained by removing the closed vertex subset $\{\mathbf{A}, \mathbf{B}, \mathbf{D}, \mathbf{E}, \mathbf{F}\}$ from the simplicial complex consisting of the two triangles **BCF**, **BDF**, the two edges **AC**, **CE**, and all their faces. The multivector field consists of four singleton edges and the doubleton $\{\mathbf{BF}, \mathbf{BDF}\}$, as well as the multivector $\{\mathbf{C}, \mathbf{BC}, \mathbf{CF}, \mathbf{BCF}\}$. The four critical edges **AC**, **BD**, **CE**, **DF** give rise to four isolated invariant sets. Furthermore, if we define the poset $P = \{\mathbf{p}, \mathbf{q}, \mathbf{r}, \mathbf{s}\}$ via the Hasse diagram

(2.3)

and the indexed family $\{M_p\}_{p \in P}$ with $M_{\mathbf{p}} := \{\mathbf{BD}\}$, $M_{\mathbf{q}} := \{\mathbf{DF}\}$, $M_{\mathbf{r}} := \{\mathbf{AC}\}$, as well as $M_{\mathbf{s}} := \{\mathbf{CE}\}$, then we obtain a well-defined Morse decomposition. Observe that all inequalities in the poset P matter, because there are connections from $M_{\mathbf{r}}$ to

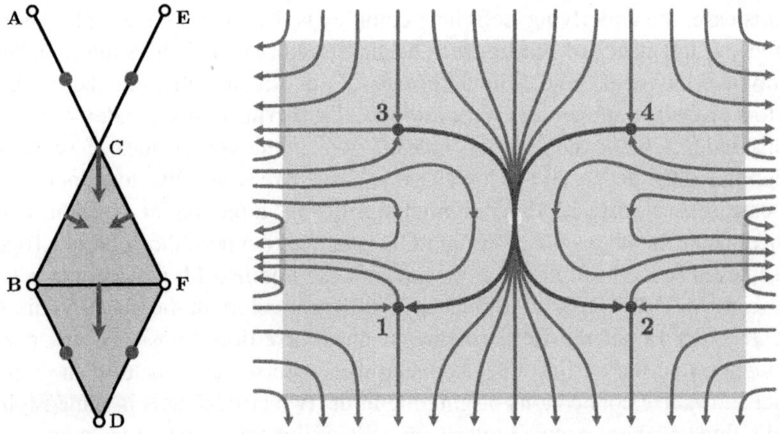

Fig. 2.3 *A multiflow without lattice of attractors.* The panel on the left shows a multivector field on a Lefschetz complex, which consists of the depicted simplicial complex, but without the vertices **A**, **B**, **D**, **E**, and **F**, shown as white circles. The multivector field consists of four singletons and one doubleton, as well as the multivector {**C**, **BC**, **CF**, **BCF**}. The dynamics of this example can be represented as the multiflow shown on the right, where the only point of forward or backward nonuniqueness is the yellow point **5**. For neither example, one can construct a lattice of attracting sets, as explained in Example 2.2.5

both $M_{\mathbf{p}}$ and $M_{\mathbf{q}}$, as well as from $M_{\mathbf{s}}$ to both $M_{\mathbf{p}}$ and $M_{\mathbf{q}}$. For instance,

$$\gamma = \overleftarrow{\mathbf{AC}} \cdot \mathbf{C} \cdot \mathbf{BCF} \cdot \mathbf{BF} \cdot \mathbf{BDF} \cdot \overrightarrow{\mathbf{BD}}$$

is an essential solution and a connection from $M_{\mathbf{r}}$ to $M_{\mathbf{p}}$. A multiflow counterpart of this example is presented in the right panel of Fig. 2.3. Its four saddle points are the stationary points marked in the figure as **1**, **2**, **3**, and **4**. The only point of nonuniqueness of solutions is the yellow point marked **5**. Similarly to the combinatorial case, there are connections through this point from the saddle **3** to both saddles **1** and **2**, as well as from saddle **4** to both saddles **1** and **2**. ◊

After this sequence of examples, we briefly return to our general discussion of Morse decompositions in combinatorial dynamical systems. Consider the family of *down sets* in the poset P, that is, the subsets $I \subset P$ such that with every $p \in I$ also all elements below p are contained in I. We denote this family as $\mathrm{Down}(P)$. It is easy to see that $\mathrm{Down}(P)$ is a *lattice of sets*, which means that it is closed under union and intersection. We now associate with every $I \in \mathrm{Down}(P)$ the set M_I consisting of all right endpoints of a path with left endpoint in the union $\bigcup_{r \in I} M_r$. One can check that the so-defined set M_I is an attractor.[4] However, the family of all

[4] The set M_I is a special instance of a *Morse interval*, which will be defined in more detail later.

such attractors is not a lattice in general, because the intersection of two attractors does not need to be an attractor itself.

In the classical setting of flows or semiflows, it is possible to overcome this difficulty by constructing an attracting neighborhood N_I for each attractor M_I in such a way that the family $\{ N_I \mid I \in \mathrm{Down}(P) \}$ is again a lattice, and such that the map $I \mapsto N_I$ is a lattice homomorphism, i.e., it preserves unions and intersections. In addition, the attracting neighborhoods N_I have to be constructed in such a way that M_I is the largest invariant set in N_I. In other words, the attracting set N_I determines the attractor M_I. In the classical setting, this lattice $\{N_I \mid I \in \mathrm{Down}(P)\}$ is then used to proceed with the construction of the connection matrix. However, as the following example indicates, such a lattice and associated lattice homomorphism may not exist in the case of a multiflow and, similarly, in the case of a combinatorial multivector field which is inherently multivalued.

Example 2.2.5 (A Multiflow Without Lattice of Attractors, $\lhd$ 2.2.4 $\rhd$ 2.3.2) Consider again the multivector field and associated multiflow shown in Fig. 2.3, which were already discussed in the previous Example 2.2.4. In both of these cases, the Morse decomposition is indexed by the same poset $P = \{\mathbf{p}, \mathbf{q}, \mathbf{r}, \mathbf{s}\}$ with Hasse diagram (2.3). One can immediately verify that the lattice of down sets of this poset is given by the collection $\{\varnothing, \{\mathbf{p}\}, \{\mathbf{q}\}, \{\mathbf{p}, \mathbf{q}\}, \{\mathbf{p}, \mathbf{q}, \mathbf{r}\}, \{\mathbf{p}, \mathbf{q}, \mathbf{s}\}, P\}$. We first observe that in the combinatorial case the set $A := X \setminus \{\mathbf{AC}, \mathbf{CE}\}$ does not admit an essential solution in A through $\mathbf{C}$, $\mathbf{BC}$, $\mathbf{CF}$, and $\mathbf{BCF}$. Therefore, A is not an isolated invariant set, and consequently also not an attractor. Notice, however, that $M_{\{\mathbf{p},\mathbf{q},\mathbf{r}\}} \cap M_{\{\mathbf{p},\mathbf{q},\mathbf{s}\}} = A$. Hence, $\{ M_I \mid I \in \mathrm{Down}(P) \}$ is not a lattice in this case.

In addition, we claim that the above-mentioned work-around for flows with the lattice of attracting neighborhoods N_I does not work in our combinatorial example either. More precisely, there exists no lattice homomorphism which sends each down set I to an attracting set N_I for the corresponding attractor, and such that M_I is the largest invariant subset of N_I. To show this, assume to the contrary that $I \mapsto N_I$ is such a homomorphism. Then one immediately obtains

$$\mathbf{C} \in M_{\{\mathbf{p},\mathbf{q},\mathbf{r}\}} \cap M_{\{\mathbf{p},\mathbf{q},\mathbf{s}\}} \subset N_{\{\mathbf{p},\mathbf{q},\mathbf{r}\}} \cap N_{\{\mathbf{p},\mathbf{q},\mathbf{s}\}} = N_{\{\mathbf{p},\mathbf{q}\}} = N_{\{\mathbf{p}\}} \cup N_{\{\mathbf{q}\}}.$$

Suppose now that the inclusion $\mathbf{C} \in N_{\{\mathbf{p}\}}$ holds. Since $\mathbf{C} \cdot \mathbf{BCF} \cdot \mathbf{BF} \cdot \mathbf{BDF} \cdot \mathbf{DF}$ is a path from $\mathbf{C}$ to $\mathbf{DF}$, and since $N_{\{\mathbf{p}\}}$ is an attracting set, this immediately yields the inclusion $\mathbf{DF} \in N_{\{\mathbf{p}\}}$ —and therefore $\{\mathbf{BD}, \mathbf{DF}\}$ is an invariant subset of $N_{\{\mathbf{p}\}}$. Yet, this is not possible, since we assumed that $M_{\{\mathbf{p}\}} = \{\mathbf{BD}\}$ is the largest invariant subset of $N_{\{\mathbf{p}\}}$. Analogously one can rule out $\mathbf{C} \in N_{\{\mathbf{q}\}}$. Altogether this indeed proves that such a lattice homomorphism does not exist. The argument for the multiflow is similar. $\Diamond$

2.3 Acyclic Partitions of Lefschetz Complexes

Before we explain how the problem with lattices of attracting sets mentioned in the last section can be addressed, we first consider the special case when the family of attractors $\{ M_I \mid I \in \mathrm{Down}(P) \}$ is indeed a lattice and $I \mapsto M_I$ is a lattice homomorphism. One can show that in the combinatorial setting an attractor is always an attracting set of itself. Thus, in this case, we also have a lattice of attracting sets—and this puts us in the setting when the approach of [21, 44] works. Actually, in view of the closedness of attracting sets, we may consider an arbitrary sublattice $\mathcal{L}$ of the lattice of closed sets in a Lefschetz complex X. Given a set $L \in \mathcal{L}$, consider now the union $L^\star$ of all proper subsets of L in $\mathcal{L}$. If the inequality $L^\star \neq L$ holds, then we call L *join-irreducible*. Clearly, the set difference

$$L^\circ := L \setminus L^\star \tag{2.4}$$

is locally closed, as a difference of two closed sets. Moreover, one can in fact prove that the family

$$\mathrm{AP}(\mathcal{L}) := \{ L^\circ \mid L \text{ is join-irreducible} \} \tag{2.5}$$

is a partition of X, which in addition is *acyclic* in the sense that the transitive closure of the relation $L_1^\circ \preceq L_2^\circ$ defined by $L_1^\circ \cap \mathrm{cl}\, L_2^\circ \neq \varnothing$ makes $\mathrm{AP}(\mathcal{L})$ into a poset. Moreover, this resulting poset turns out to be isomorphic to the poset of join-irreducibles of $\mathcal{L}$ ordered by inclusion. Actually, the presented way of passing from a lattice of sets to an acyclic partition can be viewed as a version of the celebrated Birkhoff theorem [8] for abstract lattices, yet for the special case of lattices of sets.

Example 2.3.1 (A First Multivector Field, $\lhd$ 2.2.2) For the Morse decomposition $\mathcal{M}$ discussed in Example 2.2.2, see also Fig. 2.1, the lattice of down sets is given by the collection $\{\varnothing, \{\mathbf{p}\}, \{\mathbf{p}, \mathbf{q}\}, P\}$, and one can easily see that the associated family of attractors $\mathcal{A} := \{\varnothing, \{\mathbf{D}\}, X \setminus \{\mathbf{ABC}\}, X\}$ forms a lattice. Moreover, each attractor except the empty set is join-irreducible. Thus, in this case, the corresponding acyclic partition $\mathrm{AP}(\mathcal{A})$ of X coincides with the family $\mathcal{M}$ of Morse sets, and the mapping $P \ni p \mapsto M_p \in \mathrm{AP}(\mathcal{A})$ is indeed an order isomorphism. $\Diamond$

As we explain in Sect. 2.4, and in more detail in Sect. 6.1, an acyclic partition of a Lefschetz complex may be used to define a connection matrix. Unfortunately, as we indicated in the previous section, the family $\{ M_I \mid I \in \mathrm{Down}(P) \}$ of attractors may not be a lattice in the combinatorial setting.

Therefore, in order to obtain the required acyclic partition, we have to modify the presented approach. First observe that given an arbitrary family $\mathcal{U}$ of subsets of X, there is a smallest lattice of subsets of X containing $\mathcal{U}$. We refer to it as the *lattice extension* of $\mathcal{U}$ and denote it by $\mathcal{U}'$. One can easily verify that if $\mathcal{U}$ is a finite family of closed sets, then so is its lattice extension $\mathcal{U}'$. Since, clearly, the union and intersection of attracting sets is again an attracting set, the lattice extension $\mathcal{A}'$ of the family of attractors given by $\mathcal{A} := \{ M_I \mid I \in \text{Down}(P) \}$ is a lattice of attracting sets. Hence, we can in fact construct a lattice, but it is no longer indexed by the down sets of the original poset P. Nevertheless, one can show that the acyclic partition $\mathcal{D} := \text{AP}(\mathcal{A}')$, considered as a poset, is order isomorphic to an extension $\hat{P}$ of the original poset P. This enables us to write the family $\mathcal{D}$ in the form $\mathcal{D} = \{D_p\}_{p\in\hat{P}}$ where $\hat{P} \ni p \mapsto D_p \in \mathcal{D}$ is the above-mentioned order isomorphism.

Note that every element D_p for $p \in \hat{P}$, as a locally closed subset of the Lefschetz complex X, is itself a Lefschetz complex. Moreover, one can deduce from the definition of Morse decomposition that $H(D_p) = 0$ for all $p \in \hat{P} \setminus P$. This observation will then enable us to get rid of the elements in the set difference $\hat{P} \setminus P$ via an equivalence relation introduced in the final algebraic step of our construction. Moreover, it turns out that instead of searching for the lattice extension $\mathcal{A}'$ and the associated partition, we can just take a refined acyclic partition $\mathcal{D}$ consisting of all Morse sets together with all the multivectors not contained in a Morse set. It is not difficult to see that such multivectors must be regular. Therefore, although this may lead to an acyclic partition indexed by a larger poset, the partition elements are explicitly given, and we may get rid of the elements in $\hat{P}$ coming from the regular multivectors outside Morse sets by the same algebraic equivalence. In fact, this straightforward construction replaces the laborious and generally not possible step (ii) of the connection matrix pipeline discussed in the introduction.

Example 2.3.2 (A Multiflow Without Lattice of Attractors, ◁ 2.2.5 ▷ 2.5.1) As we pointed out in Example 2.2.5, the intersection $A = M_{\{\mathbf{p},\mathbf{q},\mathbf{r}\}} \cap M_{\{\mathbf{p},\mathbf{q},\mathbf{s}\}}$ is not an attractor which implies that the family of attractors is not a lattice. Actually, in this example the set A is the only set in the lattice extension $\mathcal{A}'$ of the family of attractors $\mathcal{A}$ which is not an attractor. The set A is join-irreducible in this lattice, and therefore it adds

$$A^\circ = \{\mathbf{C}, \mathbf{BC}, \mathbf{CF}, \mathbf{BF}, \mathbf{BCF}, \mathbf{BDF}\}$$

to the associated acyclic partition $\text{AP}(\mathcal{A}')$, see also Fig. 2.3. However, the added set A° splits as the disjoint union of the two multivectors $V_1 := \{\mathbf{C}, \mathbf{BC}, \mathbf{CF}, \mathbf{BCF}\}$ and $V_2 := \{\mathbf{BF}, \mathbf{BDF}\}$. Therefore, we may take as the acyclic partition family

$$\mathcal{E} := \{ \{\mathbf{AC}\}, \{\mathbf{CE}\}, \{\mathbf{BD}\}, \{\mathbf{DF}\}, V_1, V_2 \},$$

which consists of the four singleton Morse sets and the two regular multivectors V_1 and V_2. To write it as an indexed family $\{E_p\}_{p\in\hat{P}}$, we take the extended poset $\hat{P}$ as the union $P \cup \{\mathbf{t}, \mathbf{u}\}$ with partial order given by the Hasse diagram

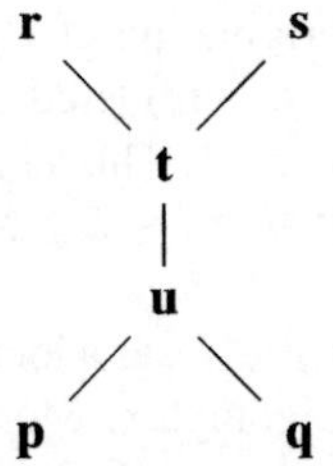

.

Moreover, we define E_p as the Morse set M_p for $p \in P$, and additionally let $E_{\mathbf{t}} := V_1$ and $E_{\mathbf{u}} := V_2$. ◊

2.4 Algebraic Connection Matrices

The acyclic partition $\{D_p\}_{p \in \hat{P}}$ of the Lefschetz complex X discussed in the previous section lets us decompose the associated chain complex $C(X)$ in the form of the direct sum $C(X) = \bigoplus_{p \in \hat{P}} C(D_p)$. It turns out that the order isomorphism $p \mapsto D_p$ gives this direct sum the special structure of a *poset filtered chain complex*, which in turn leads directly to a purely algebraic concept of connection matrix. This will be explained in more detail in the following.

We first explain the case when $\hat{P} = P$ is fixed. In terms of applications to dynamics, this corresponds to the situation when we have a lattice of attracting sets indexed by down sets of the original poset in the Morse decomposition. Recall that in this chapter we assume field coefficients for all considered modules, in particular chain complexes and homology modules. Given a chain complex (C, d) together with a direct sum decomposition

$$C = \bigoplus_{p \in P} C_p \tag{2.6}$$

and a down set $I \in \mathrm{Down}(P)$, we introduce the abbreviation $C_I := \bigoplus_{p \in I} C_p$. We say that the boundary homomorphism d is *filtered* if $d(C_I) \subset C_I$ is satisfied for every $I \in \mathrm{Down}(P)$. If this is the case, then the chain complex (C, d) together with the decomposition (2.6) is called a *P-filtered chain complex*. Similarly, any homomorphism, and in particular, any chain map $h : C \to C'$ between two P-filtered chain complexes, is called *filtered* if

$$h(C_I) \subset C'_I \quad \text{for every} \quad I \in \mathrm{Down}(P). \tag{2.7}$$

We would like to point out that every acyclic partition $\mathcal{E}$ of a Lefschetz complex X makes $C(X)$, the chain complex of X, a filtered chain complex via the decomposition

$$C(X) = \bigoplus_{E \in \mathcal{E}} C(E).$$

The decomposition is well-defined, because each $E \in \mathcal{E}$, as a locally closed subset of X, is itself a Lefschetz complex, and the fact that the boundary homomorphism is filtered may be concluded from the assumption that the partition is acyclic, see Proposition 6.1.3.

Now, two P-filtered chain complexes are *filtered chain homotopic* if there exist filtered chain maps $h : C \to C'$ and $h' : C' \to C$ such that the composition $h'h$ is filtered chain homotopic to id_C, and hh' is filtered chain homotopic to $\mathrm{id}_{C'}$. In this context, a *filtered chain homotopy* is a chain homotopy which is itself filtered as a homomorphism. Robbin and Salamon [44] prove that every P-filtered chain complex is P-filtered chain homotopic to a *reduced* chain complex, that is, a filtered chain complex (C', d') such that $d'_{pp} = 0$ for all $p \in P$. Moreover, Harker et al. [21] prove that if (C, d) is also filtered chain homotopic to another reduced filtered complex (C'', d''), then the filtered chain complexes (C', d') and (C'', d'') are in fact filtered isomorphic. Hence, up to filtered chain isomorphism, every filtered chain complex has exactly one reduced representation. By definition, this representation is called its *Conley complex*, and the matrix of the boundary operator of the Conley complex is referred to as its *connection matrix*.

Example 2.4.1 (Small Lefschetz Complex with Periodic Orbit, ▷ 2.5.2) This example lays the groundwork for the discussion of a combinatorial dynamical system on a small Lefschetz complex which has a periodic orbit. While the more geometric description is delayed until Example 2.5.2, we begin by considering the algebra behind it, in a purely abstract setting. For this, consider free module $C = R\langle X \rangle$ spanned by the set of symbols

$$X := \{\mathbf{A}, \mathbf{B}, \mathbf{a}, \mathbf{b}, \mathbf{c}, \boldsymbol{\alpha}\}$$

and with partition $X = X_0 \cup X_1 \cup X_2$ given by

$$X_0 := \{\mathbf{A}, \mathbf{B}\}, \quad X_1 := \{\mathbf{a}, \mathbf{b}, \mathbf{c}\}, \quad X_2 := \{\boldsymbol{\alpha}\}.$$

This partition allows us to treat the free module C as a $\mathbb{Z}$-graded module with gradation $C = \bigoplus_{i=0}^{2} R\langle X_i \rangle$. In order to make C a free chain complex, we assume that R is the field $\mathbb{Z}_2$ and consider a homomorphism $d : C \to C$ defined on the basis X by the matrix

d	**A**	**B**	**a**	**c**	**b**	$\boldsymbol{\alpha}$
A			1	1	1	
B			1	1	1	
a						1
c						
b						1
$\boldsymbol{\alpha}$						

It is not difficult to verify that the pair (C, d) is indeed a free chain complex. Furthermore, it can be turned into a poset filtered chain complex by considering the poset $P := \{\mathbf{p}, \mathbf{q}, \mathbf{r}\}$, linearly ordered by $\mathbf{p} < \mathbf{q} < \mathbf{r}$, and by setting

$$C_{\mathbf{p}} := R\langle\{\mathbf{A}, \mathbf{B}, \mathbf{a}, \mathbf{c}\}\rangle, \quad C_{\mathbf{q}} := R\langle\{\mathbf{b}\}\rangle, \quad C_{\mathbf{r}} := R\langle\{\boldsymbol{\alpha}\}\rangle.$$

Again, one can easily check that the gradation $C = \bigoplus_{p \in P} C_p$ turns the pair (C, d) into a P-filtered chain complex. Since we have $d_{\mathbf{pp}} \neq 0$, this complex C is not reduced, and therefore not a Conley complex.

One can also consider a submodule $\bar{C}$ of C which is obtained by removing the generators **B** and **c**, i.e., we have

$$\bar{C}_{\mathbf{p}} := R\langle\{\mathbf{A}, \mathbf{a}\}\rangle, \quad \bar{C}_{\mathbf{q}} := R\langle\{\mathbf{b}\}\rangle, \quad \bar{C}_{\mathbf{r}} := R\langle\{\boldsymbol{\alpha}\}\rangle,$$

together with the boundary homomorphism $\bar{d} : \bar{C} \to \bar{C}$ given by the matrix

$\bar{d}$	**A**	**a**	**b**	$\boldsymbol{\alpha}$
A				
a				1
b				1
$\boldsymbol{\alpha}$				

We leave it to the reader to verify that this defines a P-filtered chain complex $(\bar{C}, \bar{d})$. Furthermore, this chain complex is reduced and, in fact, is a Conley complex of (C, d) which makes the matrix of $\bar{d}$ a connection matrix of (C, d). To see this in more detail, consider the two specific homomorphisms $h : C \to \bar{C}$ and $g : \bar{C} \to C$

given by the matrices

h	**A**	**B**	**a**	**c**	**b**	α
A	1	1				
a			1			
b					1	
α						1

and

g	**A**	**a**	**b**	α
A	1			
B				
a		1		
c		1	1	
b			1	
α				1

.

One can immediately verify that both g and h are P-filtered chain maps and that the identity $h \circ g = \mathrm{id}_{\bar{C}}$ is satisfied. Moreover, the composition $g \circ h$ can be computed as the matrix

$g \circ h$	**A**	**B**	**a**	**c**	**b**	α
A	1	1				
B						
a			1			
c			1		1	
b					1	
α						1

.

Therefore, this composition $g \circ h$ is filtered chain homotopic to id_C via the P-filtered chain homotopy $\gamma : C \to C$ which maps all generators to zero, except **B** which is mapped to **c**. In fact, this chain homotopy is an example of an *elementary reduction* via the pair of generators $(\mathbf{B}, \mathbf{c})$ discussed in [23, Section 4.3], see also [24]. Elementary reductions provide a tool to compute Conley complexes and connection matrices by hand in simple problems. We presented this example purely algebraically, but as stated above, its algebra provides the connection matrix for the Morse decomposition of a combinatorial multivector field which we will discuss in more detail in Example 2.5.2. $\Diamond$

As we mentioned already earlier, one has to extend the definitions of Conley complex and connection matrix discussed so far to accommodate the situation when we cannot keep the original poset P as in Examples 2.2.4, 2.2.5, and 2.3.2. Under a changing poset, the condition (2.7) in the definition of a filtered homomorphism needs to be modified. For this, let P and P' be two posets, and consider a P-filtered chain complex (C, d), as well as a P'-filtered chain complex (C', d'). In order to speak about a filtered homomorphism in this setting, we need a mechanism to relate down sets in the two posets P and P'. Note that if the map $\alpha : P' \to P$ is order preserving, then we have $\alpha^{-1}(I) \in \mathrm{Down}(P')$ for every $I \in \mathrm{Down}(P)$. Therefore, one can define a *morphism* from a P-filtered chain complex (C, d) to a P'-filtered chain complex (C', d') as a pair (α, h), where $\alpha : P' \to P$ is an order preserving mapping and $h : C \to C'$ is a chain map satisfying

$$h(C_I) \subset C'_{\alpha^{-1}(I)} \quad \text{for every} \quad I \in \mathrm{Down}(P). \tag{2.8}$$

A similar modification allows us to also extend the definition of filtered chain homotopy to the setting of varying posets.

We still need some further modifications to guarantee that in this algebraic step we can eliminate the elements added to the poset to obtain a lattice of attracting sets. For this, we recall from Sect. 2.3 that subcomplexes C_p for the added values of p are Lefschetz complexes with zero homology. Equivalently, they are chain homotopic to zero (see Corollary 3.4.7). Therefore, if $\bar{C}$ denotes the Conley complex of C computed under a fixed poset P, the chain groups $\bar{C}_p$ for an added element p are zero. In consequence, the respective rows and columns in the connection matrix are empty, because the only basis of the zero group is empty. To formalize the removal of these empty rows and columns, we do four things:

- We designate a distinguished subset $P_\star \subset P$ and require that all chain complexes C_p for the elements $p \in P \setminus P_\star$ are chain homotopic to zero. One can think of the elements in $P_\star$ as significant, and the ones in the complement $P \setminus P_\star$ as insignificant. We assume $H(C_p) \neq 0$ to be a sufficient condition for $p \in P$ to be significant. However, the condition need not be necessary, because there may be other reasons related to a concrete application to consider a $p \in P$ significant.
- To remove the insignificant elements from the Conley complex and connection matrix, we extend the definition of a reduced chain complex by additionally requesting that $P_\star = P$ in a reduced complex.
- We allow partial order preserving maps $\alpha : P' \nrightarrow P$ to relate down sets in the posets P and P', but we require that α is defined at least for all $p' \in P'_\star$. The partial map α does not need to be defined for elements $p' \in P' \setminus P'_\star$, since the corresponding term $\bar{C}'_p = 0$ in the Conley complex contributes nothing to the whole Conley complex.
- Finally, since the preimage $\alpha^{-1}(I)$ for $I \in \mathrm{Down}(P)$ no longer needs to be a down set under a partially defined order preserving map, we also have to replace (2.8) by the condition

  $$h(C_I) \subset C'_{\alpha^{-1}(I)^{\leq}} \quad \text{for every} \quad I \in \mathrm{Down}(P), \tag{2.9}$$

  where $A^{\leq}$ denotes the smallest down set containing A, for $A \subset P'$. We refer to chain maps satisfying (2.9) as *α-filtered*.[5]

As will become clear in the next section, in applications to dynamics, we generally start with a poset P, we then extend it to a poset $\hat{P}$ to resolve some technical issues like the problems discussed in Examples 2.2.5 and 2.3.2, and finally we apply the

[5] In fact, when we formally define this concept later, we instead use the definition (4.2) for technical reasons. Its equivalence with (2.9) is established in Proposition 4.1.3.

algebraic connection matrix theory to the extended poset $\hat{P}$ with the distinguished subset $(\hat{P})_\star$ being equal to the original poset P.

Summarizing, the above extensions lead to a well-defined category PFCC, constructed in detail in Chap. 4. The objects of this category are of the form (P, C, d), where P is a poset with a distinguished subset $P_\star$ and (C, d) is a P-filtered chain complex. In addition, morphisms from (P, C, d) to (P', C', d') are of the form (α, h), where $\alpha : P' \nrightarrow P$ is a partial order preserving map whose domain contains $P'_\star$ and which satisfies $\alpha(P'_\star) \subset P_\star$, and $h : C \to C'$ is a chain map satisfying (2.9). As it turns out, the main results of [21, 44] can be extended to this new category. We do this in detail in Chap. 5 (see, in particular, Corollary 5.2.6 and Theorem 5.3.2). The results of Chap. 5 can be summarized as the following theorem.

Theorem 2.4.2 (Existence of the Conley Complex and Connection Matrix) *Consider the category* PFCC *of poset filtered chain complexes with varying posets introduced above. Then the following hold:*

(i) Every object in PFCC *is filtered chain homotopic to a reduced object.*
(ii) If two reduced objects in PFCC *are filtered chain homotopic, then they automatically are isomorphic in* PFCC.

In other words, up to isomorphism, every object in PFCC *has exactly one reduced representation, which is called its* Conley complex. *The matrix of the boundary operator of the Conley complex is called* connection matrix. □

2.5 Connection Matrices of Morse Decompositions

In the previous section, we introduced both the Conley complex and the connection matrix of a filtered chain complex. We also defined the Conley complex and the connection matrix of an acyclic partition of a Lefschetz complex via the filtered chain complex induced by such a partition. We now turn our attention to connection matrices of Morse decompositions.

Hence, assume that we are given a Morse decomposition $\mathcal{M} = \{M_p\}_{p \in P}$ of a combinatorial multivector field $\mathcal{V}$ on a Lefschetz complex X. We want to define a connection matrix of $\mathcal{M}$ via an acyclic partition associated with $\mathcal{M}$. By definition, the collection $\mathcal{V}$ is a partition of X, but it may not be acyclic. However, by merging multivectors inside the same Morse set, we do get an acyclic partition. It is defined as the family $\{E_p\}_{p \in \hat{P}}$ where $\hat{P}$ is an extension of P such that $E_p = M_p$ for $p \in P$, and each multivector V not contained in a Morse set equals E_q for a unique $q \in \hat{P} \setminus P$ (see Definition 7.2.9). As we prove in Corollary 7.2.8, every E_q for $q \in \hat{P} \setminus P$ is a regular multivector, and therefore we can take $(\hat{P})_\star := P$ in this construction. This means, via Theorem 2.4.2, that the Conley complex and the connection matrix of $\mathcal{M}$ defined as the Conley complex and connection matrix of $\mathcal{E}_{\mathcal{M}} := \{E_p\}_{p \in \hat{P}}$ bring us back to the original poset P.

The discussed construction provides both the Conley complex and the connection matrix also in the case when there is no lattice homomorphism from down sets in the poset of a Morse decomposition to attracting sets in this Morse decomposition, as discussed earlier. A bonus, which comes as a side effect, is that we can take a shortcut in the connection matrix pipeline by skipping the construction of the lattice of attracting neighborhoods and passing immediately from the Morse decomposition $\mathcal{M}$ in step (i) to a filtered chain complex in step (iii) via the acyclic partition $\mathcal{E}_{\mathcal{M}}$ of the phase space associated with the Morse decomposition $\mathcal{M}$ under the extended poset.

Example 2.5.1 (A Multiflow Without Lattice of Attractors, $\lhd$ 2.3.2) Consider the acyclic partition $\mathcal{E} = \{E_p\}_{p\in\hat{P}}$ introduced in Example 2.3.2, which corresponds to the Morse decomposition of the multivector field on the Lefschetz complex X in the left panel of Fig. 2.3. The boundary homomorphism d of the associated chain complex $C(X)$ of X has the matrix

d	**BD**	**DF**	**BF**	**BDF**	**C**	**BC**	**CF**	**BCF**	**AC**	**CE**
BD				1						
DF				1						
BF				1				1		
BDF										
C						1	1		1	1
BC								1		
CF								1		
BCF										
AC										
CE										

.

One can immediately verify that $(C(X), d)$ defines indeed a $\hat{P}$-filtered chain complex. For $p \in \hat{P} \setminus P = \{\mathbf{t}, \mathbf{u}\}$ the induced Lefschetz complex X_p is chain homotopic to zero. Therefore, we take $\hat{P}_* := P$ as the distinguished subset in $\hat{P}$. This way one obtains an object $(\hat{P}, C(X), d)$ of PFCC. Notice that d is not reduced, because $d_{\mathbf{tt}} \neq 0$, $d_{\mathbf{uu}} \neq 0$, and additionally, $C(X_p)$ is chain homotopic to zero for $p \in \hat{P} \setminus P$.

Consider now the free module $R\langle\mathbf{BD}, \mathbf{DF}, \mathbf{AC}, \mathbf{CE}\rangle$ and turn it into a chain complex $\bar{C}$ by letting the boundary be identically zero. Then the definitions $\bar{C}_{\mathbf{p}} := \langle\mathbf{BD}\rangle$, $\bar{C}_{\mathbf{q}} := \langle\mathbf{DF}\rangle$, $\bar{C}_{\mathbf{r}} := \langle\mathbf{AC}\rangle$, and $\bar{C}_{\mathbf{s}} := \langle\mathbf{CE}\rangle$ make $\bar{C}$ a P-filtered chain complex. After finally setting $\bar{P} := \bar{P}_\star := P$, we now claim that $(\bar{P}, \bar{C}, 0)$ is a Conley complex of the Morse decomposition associated with the multivector field on the Lefschetz complex X in the left panel of Fig. 2.3. In order to see this, let $\alpha : \bar{P} \hookrightarrow \hat{P}$ denote the inclusion map, and let $\beta : \hat{P} \nrightarrow \bar{P}$ be the partial map which, as a relation, is the inverse of α. Consider the homomorphisms $h : C(X) \to \bar{C}$ and $g : \bar{C} \to C(X)$ given by the matrices

h	**BD**	**DF**	**BF**	**BDF**	**C**	**BC**	**CF**	**BCF**	**AC**	**CE**
BD	1		1				1			
DF		1	1				1			
AC									1	
CE										1

and

g	**BD**	**DF**	**AC**	**CE**
BD	1			
DF		1		
BF				
BDF				
C				
BC			1	1
CF				
BCF				
AC			1	
CE				1

.

One can readily verify that the mapping h is an α-filtered chain map and that g is a β-filtered chain map. Hence, both (α, h) and (β, g) are morphisms in PFCc. It turns out that $(\alpha, h) \circ (\beta, g) = \mathrm{id}_{(\bar{P}, \bar{C}, 0)}$ and that $(\beta, g) \circ (\alpha, h)$ is filtered chain homotopic to $\mathrm{id}_{(\hat{P}, C(X), d)}$ via the $\hat{P}$-filtered homotopy $\gamma : C(X) \to C(X)$ defined as

γ	**BD**	**DF**	**BF**	**BDF**	**C**	**BC**	**CF**	**BCF**	**AC**	**CE**
BD										
DF										
BF										
BDF			1				1			
C										
BC					1					
CF										
BCF							1			
AC										
CE										

.

This confirms that $(\bar{P}, \bar{C}, 0)$ is in fact a Conley complex of the Morse decomposition associated with the multivector field on the Lefschetz complex X in the left panel of Fig. 2.3. ◊

We have already presented one example of a connection matrix in the context of Example 2.2.2. In that case, it could easily be derived from the fact that the homology of the Conley complex is isomorphic to the homology of the underlying Lefschetz complex. A more complicated example will further illustrate Theorem 2.4.2.

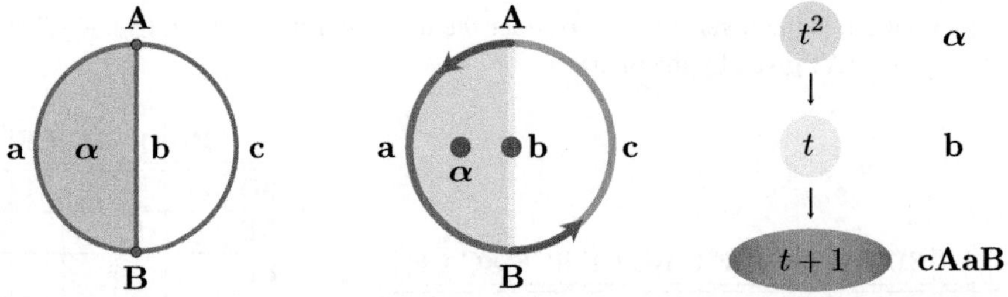

Fig. 2.4 *Small Lefschetz complex with periodic orbit.* The left panel shows a small Lefschetz complex which consists of a 2-cell $\boldsymbol{\alpha}$, three 1-cells **a**, **b**, **c**, and two 0-cells **A**, **B**. On this complex, we study the combinatorial vector field shown in the middle panel, which consists of two singletons and two doubletons. The associated Conley-Morse graph is shown on the right, with Morse sets given by the critical cells $\{\boldsymbol{\alpha}\}$ and $\{\mathbf{b}\}$, as well as the periodic orbit $\{\mathbf{c}, \mathbf{A}, \mathbf{a}, \mathbf{B}\}$

Example 2.5.2 (Small Lefschetz Complex with Periodic Orbit, ◁ 2.4.1) As a more elaborate example, yet one that can still be discussed directly in detail, consider the Lefschetz complex X shown in the left panel of Fig. 2.4. This complex consists of a semicircle shaped two-dimensional cell $\boldsymbol{\alpha}$, whose boundary consists of the vertices **A** and **B**, together with the two 1-cells **a** and **b**. In addition, the complex contains a third 1-cell **c** which joins the two vertices. On X, we consider the combinatorial vector field $\mathcal{V}$ sketched in the middle panel of Fig. 2.4 and which consists of two singletons and two doubletons. The associated Conley-Morse graph is shown on the right, with Morse sets given by the critical cells $\{\boldsymbol{\alpha}\}$ and $\{\mathbf{b}\}$, as well as the periodic solution $\{\mathbf{c}, \mathbf{A}, \mathbf{a}, \mathbf{B}\}$. Due to the small size of the Lefschetz complex X, one can immediately reduce the associated chain complex via elementary reduction pairs. For example, if one uses the reduction pair (B, c), one obtains the sequence of Lefschetz complexes shown in the top arrow sequence in Fig. 2.5, and this leads to the connection matrix shown in Table 2.1(left). We discussed algebraic details of the computation of this connection matrix in Example 2.4.1. In contrast, if one uses the reduction pair (A, a), then the sequence of Lefschetz complexes is as in the bottom arrow sequence of Fig. 2.5, and one obtains the connection matrix in Table 2.1(right). It turns out that the matrices in Table 2.1 are not equivalent in the sense briefly introduced in Sect. 2.6 and discussed in detail in Sect. 5.5. The algebraic details of the computations are similar to those presented in Example 2.4.1 and Example 5.5.7. ◊

2.6 Uniqueness of Connection Matrices

Theorem 2.4.2(ii) implies that connection matrices are uniquely determined up to an isomorphism in PFCc. While this statement is clearly correct, it does not convey the complete story. It was observed in [20] and [41, 42] that a certain nonuniqueness

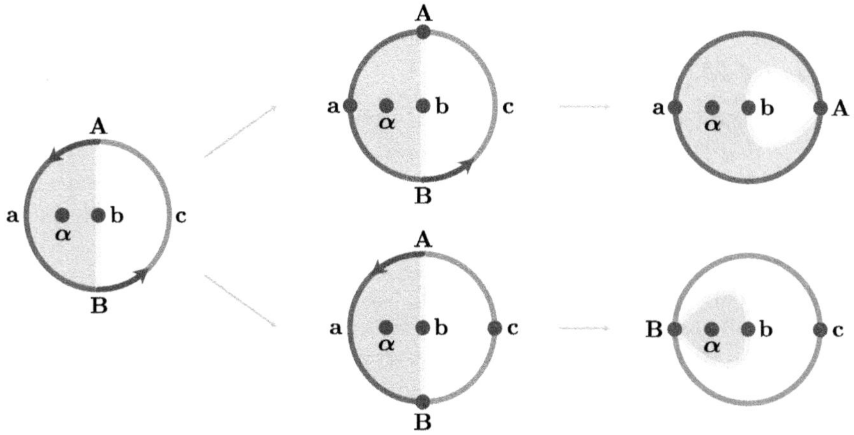

Fig. 2.5 *Small Lefschetz complex with periodic orbit.* For the combinatorial vector field shown in Fig. 2.4, the above diagram sketches the computation of two different connection matrices, based on two different elementary chain complex reductions. The top arrow sequence uses the reduction pair (B, c), while the bottom sequence is for (A, a). In both cases, the final chain complex is reduced and boundaryless

Table 2.1 *Small Lefschetz complex with periodic orbit.* Two connection matrices for the combinatorial vector field from Fig. 2.4. The matrix on the left is the result of the reduction process indicated via the top arrows in Fig. 2.5, while the bottom arrows lead to the right matrix

	A	a	b	α
A		0	0	
a				1
b				1
α				

	B	c	b	α
B		0	0	
c				0
b				1
α				

of connection matrices, observed in dynamical terms, reflects bifurcations in the underlying dynamics. Thus, it is natural to expect some form of nonuniqueness also in a purely algebraic setting. The uniqueness up to an isomorphism in PFCc only means that any two connection matrices of a given filtered chain complex are similar via the matrix of a filtered chain map. However, there is also a stronger equivalence relation between connection matrices, under which connection matrices need not be unique.

To explain this, recall that every Conley complex of a given filtered chain complex, considered as an object of PFCc, may be considered together with the filtered chain maps establishing the filtered chain homotopy. The composition of these filtered chain maps between the two Conley complexes of a given filtered chain complex is a filtered chain map. Nevertheless, this composition may or may not be chain homotopic to a *graded chain map*, which we define as a chain map $h : C_p \to C'_p$ such that $h(C_p) \subset C'_{\alpha^{-1}(p)}$. This leads to a stronger equivalence relation between connection matrices which lets us speak about uniqueness and nonuniqueness of connection matrices. In particular, one of the main new results of the book is the following theorem, whose precise formulation can be found in Theorem 8.4.5.

Theorem 2.6.1 (Unique Connection Matrix for Gradient Vector Fields) *Assume that $\mathcal{V}$ is a gradient combinatorial vector field on a regular Lefschetz complex X. Then the Morse decomposition consisting of all the critical cells of $\mathcal{V}$ has precisely one connection matrix. It coincides with the matrix of the boundary operator of the associated Conley complex.* □

In addition to being unique, the connection matrix of a gradient combinatorial vector field also allows us to establish connections between Morse sets, as the following result shows. Its precise formulation is the subject of Theorem 8.4.6, and this result also explains how the connections can be found explicitly.

Theorem 2.6.2 (Existence of Connecting Orbits) *Assume that $\mathcal{V}$ is a gradient combinatorial vector field on a regular Lefschetz complex X, and let M_p and M_q with $p < q$ denote two critical cells of $\mathcal{V}$ such that the entry in the connection matrix corresponding to these cells is nonzero. Then there exists a connection between the critical cells M_q and M_p.* □

In fact, the above result under slightly stronger assumptions remains true for general multivector fields, see Theorem 7.3.6. We illustrate these two results in the following example.

Example 2.6.3 (Three Forman Gradient Vector Fields, ▷ 8.1.5) Consider the three combinatorial Forman vector fields shown in the left column of Fig. 2.6. All of them are defined on the same simplicial complex, and they consist of eight critical cells and seven doubletons. In fact, one can easily see that all of these vector fields are related to the combinatorial vector field from Fig. 2.2, which has a periodic orbit traversing the vertices **C**, **D**, and **E**. While the latter vector field is therefore not gradient, it is possible to destroy its recurrent behavior by replacing precisely one vector in the periodic orbit by two critical cells—one edge and one vertex. Since the periodic orbit consists of three vectors, this leads to the three different gradient vector fields shown in the left column of Fig. 2.6. Their corresponding Morse decompositions are depicted in Fig. 2.7.

Using our results from Chap. 8, one can readily determine the unique connection matrix of a combinatorial gradient vector field. This basically amounts to finding suitable bases of chains in each dimension and creating the matrix of the boundary operator with respect to these new bases, see the detailed description in Theorem 8.4.5 and Proposition 8.3.2. For the gradient combinatorial vector field in the last row of Fig. 2.6, this is illustrated in detail in Example 8.4.7, and this leads to the last connection matrix in Table 2.2. For the first two gradient vector fields, their connection matrices are given by the first two matrices in Table 2.2, respectively, and they can be determined using the basis elements described in Example 8.2.4. Notice that the nonzero entries in all three of these matrices between critical cells of index 2 and critical cells of index 1 yield precisely the connecting orbits shown in the right column of Fig. 2.6. Similarly, nonzero entries between critical cells of index 1 and

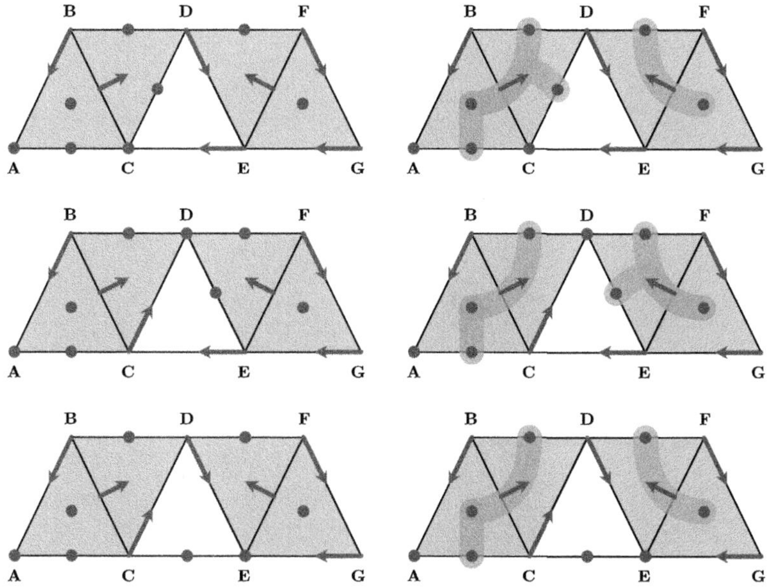

Fig. 2.6 *Three Forman gradient vector fields*. The above panels show three different combinatorial Forman gradient vector fields. All of them are related to the combinatorial vector field from Fig. 2.2, yet they destroy its periodic orbit which traverses the vertices **C**, **D**, and **E**. This is achieved by replacing precisely one vector in the periodic orbit by two critical cells—one edge and one vertex. Since the periodic orbit consists of three vectors, there are three different ways of breaking it, leading to the vector fields in the left column. In the right column, we indicate for each case which connections exist between the index 2 and index 1 critical cells

critical cells of index 0 do correspond to connections in the combinatorial vector fields. Notice, however, that the entry in the connection matrix records also the multiplicity of such connections. This is the reason why the connection matrices record the connections between **BD** or **AC** and the vertex **A**, as well as with the additional critical vertex in the support of the destroyed periodic orbit. In contrast, connections originating at either **DF** or the critical cell of dimension one on the former periodic orbit do not give rise to nonzero entries in the connection matrix, since in these cases the two different connections end at the same critical vertex. ◊

Example 2.6.4 (A Forman Vector Field with Periodic Orbit, ◁ 2.2.3 ▷ 8.4.8) Based on the previous example, we can now revisit the combinatorial vector field from Fig. 2.2, which has already been discussed in Example 2.2.3. Notice that this vector field has a periodic orbit which traverses the vertices **C**, **D**, and **E**. The existence of this periodic orbit makes the system non-gradient, and therefore we cannot apply Theorem 2.6.1 to obtain the existence of a unique connection matrix.

Nevertheless, Theorem 2.6.1 guarantees the existence of a connection matrix in this situation as well. In order to explain how such a matrix can be determined using the previous example, note that the Conley polynomial of the periodic orbit is given

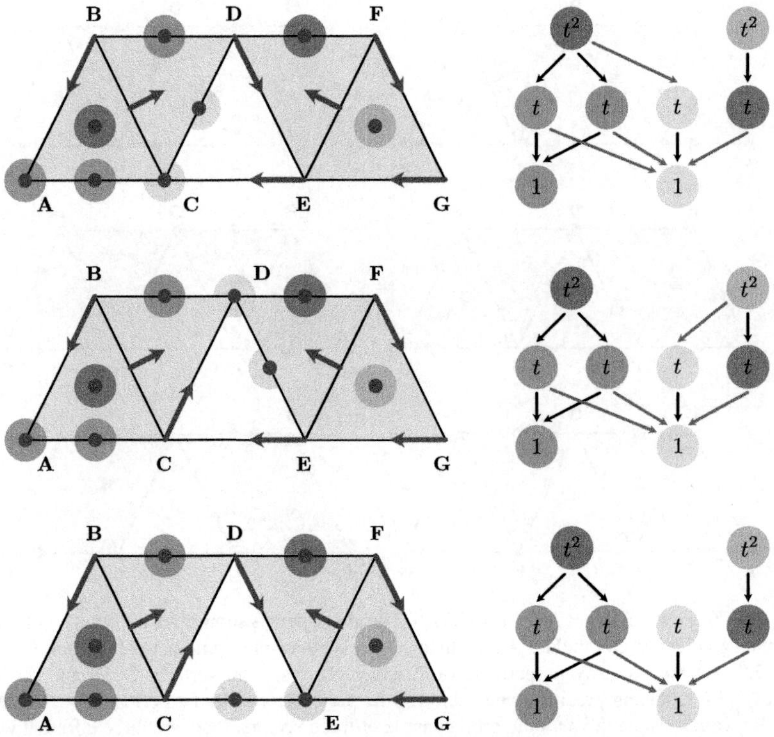

Fig. 2.7 *Three Forman gradient vector fields*. For the combinatorial vector fields shown in Fig. 2.6, the above panels depict their associated Morse decompositions. The Morse sets are indicated by different colors in the left column, while the Conley-Morse graphs are shown on the right. In each case, the two Morse sets obtained via breaking the periodic orbit in Fig. 2.2 are shown in yellow

by $t+1$, i.e., the associated homology groups are one-dimensional in dimensions 0 and 1. In Example 2.6.3, we saw that by replacing a vector on the periodic orbit by one critical cell each of dimension 0 and dimension 1, one obtains a gradient system with a unique connection matrix. In fact, the sum of the homology groups of these newly added critical cells is isomorphic to the Conley index of the periodic orbit, and therefore one would expect that all three connection matrices from Table 2.2 should be connection matrices for the vector field in Fig. 2.2. As we will see later, this is indeed the case, and all three matrices are not equivalent. Moreover, in all three cases, the nontrivial entries correspond to connections between the respective Morse sets. For a more detailed explanation, we refer the reader to Example 8.4.8. ◊

Table 2.2 *Three Forman gradient vector fields*. For the three combinatorial vector fields shown in Fig. 2.6 and associated Morse decompositions in Fig. 2.7, the three unique connection matrices are listed on the right. Only the two right-most entries in the third row are changing, and they correspond to the connections from the index 2 critical cells to the index 1 critical cell on the destroyed periodic orbit. Note that the nonzero entries in the lower right 4×2-submatrices correspond exactly to the connections indicated in the right column of Fig. 2.6

	CD	AC	BD	DF	ABC	EFG
A	0	1	1	0		
C	0	1	1	0		
CD					1	0
AC					1	0
BD					1	0
DF					0	1

	DE	AC	BD	DF	ABC	EFG
A	0	1	1	0		
D	0	1	1	0		
DE					0	1
AC					1	0
BD					1	0
DF					0	1

	CE	AC	BD	DF	ABC	EFG
A	0	1	1	0		
E	0	1	1	0		
CE					0	0
AC					1	0
BD					1	0
DF					0	1

2.7 Computability of Connection Matrices

While the last two sections have introduced the notions of Conley complex and connection matrix, so far we have only discussed small examples that could be worked out in detail. Yet, even in those cases, we have not presented a systematic method which allows one to compute connection matrices in general. As it turns out, up until very recently, there were no direct algorithms available for deriving connection matrices. In most applications, connection matrices were determined in an indirect way, by combining their properties with independently obtained Morse decomposition and Conley index information to narrow down the possibilities—often until only one final choice was left. In such a case, in view of the abstract existence result by Franzosa [20], one had clearly found the connection matrix. Despite its haphazardness, this indirect approach proved to be useful in many cases. For an example involving parabolic partial differential equations, see [31], which also cites numerous other results of this type.

Beginning with the work [21, 46] by Spendlove and collaborators, this significant gap in the theory of connection matrices has been closed. At the time of writing, there are two different approaches available:

- The original algorithm described in [21, 46] is based on ideas from discrete Morse theory. It introduces reduction techniques that can be used to transform the input complex in several sweeps to smaller complexes, and it will terminate with a Conley complex and associated connection matrix. Different connection matrices can potentially be obtained by imposing different admissible orders on the cells in the initial cell complex.
- The second approach was introduced in [13], and it is motivated by the standard algorithm to compute persistence as described for example in [15]. It works directly on the boundary matrix of the Lefschetz complex, after a reordering of the columns and rows which takes into account the underlying multivector structure and the resulting Morse decomposition. Also in this algorithm, different reorderings can lead to different connection matrices.

For the purposes of this book, we make use of a recent implementation of the second approach in the open-source mathematical programming language Julia [7]. The package `ConleyDynamics.jl` provides an implementation of the persistence-like algorithm for multivector fields on general Lefschetz complexes, which can be considered over arbitrary finite fields or over the rationals $\mathbb{Q}$. For more details on the package and its usage, we refer the reader to [48, 49]. It will be used for all of the illustrations in the next section of the chapter. In fact, the package provides predefined functions which generate the Lefschetz complexes and associated multivector fields for all of the examples that can be found in this book.

2.8 From Combinatorial Dynamics to Classical Flows

Over the last few sections, we have demonstrated how the global dynamics of multivector fields on Lefschetz complexes can be analyzed using Conley theory and connection matrices. It is natural to wonder whether these combinatorial considerations have any bearing on the case of classical dynamics on Euclidean spaces. In fact, one can ask this question in two directions. On the one hand, do multivector fields correspond to similar classical dynamical systems on a geometric realization of the Lefschetz complex? On the other hand, is it possible to leverage the language of multivector fields to make statements about the global behavior of concrete classical dynamical systems such as flows?

The first of these questions has partially been answered by Mrozek and Wanner [38]. In this paper, it is shown that for every combinatorial Forman vector field $\mathcal{V}$ on an arbitrary simplicial complex X, there exists a classical continuous semiflow φ on the underlying geometric realization $|X|$ of the complex which exibits the same dynamics as $\mathcal{V}$. More precisely, there exists a Morse decomposition for φ which has the same Conley-Morse graph as the combinatorial vector field $\mathcal{V}$. The semiflow φ is constructed explicitly through an ordinary differential equation involving the barycentric coordinates associated with the simplicial complex. In fact, the paper also introduces a notion of *admissible flow* which is based on transversality considerations, and it is shown that every such admissible flow has the same dynamics as $\mathcal{V}$. While this result only considers the case of Forman vector fields and simplicial complexes, it does show that the combinatorial theory described above can be used to easily construct classical dynamical systems with prescribed dynamical behavior.

For the remainder of this section, we turn our attention to the second question. Can multivector fields be used to understand the global dynamics of classical dynamical systems? It is a well-known fact that even in low dimensions, nonlinear ordinary differential equations cannot be solved explicitly in general. Thus, their dynamics can only be understood through the application of *qualitative methods*, which infer the solution behavior in an indirect way. Among approaches of this type, the ones relying on topology, such as Conley theory, are of particular interest, since they usually lead to results which are robust to small perturbations. This in turn paves the way for the use of computer-assisted verification techniques.

One intuitive method for obtaining rigorous qualitative results about low-dimensional dynamics in a computer-assisted way can be described as follows. Suppose we are given a classical flow φ which is generated by a planar ordinary differential equation $\dot{x} = f(x)$, where $x \in U \subset \mathbb{R}^2$. Suppose further that the domain U has been triangulated in such a way that along every triangle edge of the decomposition, the flow φ is *transverse*, in the sense that the dot product of the vector field f and a normal vector to the edge is never zero. If this is the case, then between any two adjacent triangles in the decomposition, there is a unique direction of the flow. Thus, one should be able to find isolating neighborhoods and index pairs which consist of unions of triangles, and which can then be used to prove the existence of certain isolated invariant sets. For example, in the context of establishing periodic solutions and chaos, results of this type were obtained in [37]. Moreover, at least in principle, this approach should also be usable in small dimensions larger than two.

While the above method seems promising at first glance, in practice it is extremely difficult to construct a suitable decomposition into triangles where the flow across *every face* is transverse. The whole process would be considerably simpler if the decomposition could consist of more general regions, and if flow transversality would not have to be enforced across every face. A first step toward such a relaxation has been made in [9, 16]. However, as we demonstrate briefly in the following, a wide-ranging and systematic extension can be achieved through the use of combinatorial multivector fields [48, 50]. To explain this, we begin by quoting the following simple result from [48].

Theorem 2.8.1 (Minimal Multivector Fields via Dynamical Transitions) *Let X denote a Lefschetz complex and let $\mathcal{D}$ denote an arbitrary collection of subsets of X. Then there exists a uniquely determined minimal multivector field $\mathcal{V}$ which has the property that for every set $D \in \mathcal{D}$ there exists a $V \in \mathcal{V}$ with $D \subset V$.* □

In the above result, one should think of $\mathcal{D}$ as the collection of all allowable *dynamical transitions* that can potentially occur between cells in X, in addition to the always-allowed flow toward the boundary of a cell. This leads to a less restrictive approach to discretizing the phase space of a low-dimensional ordinary differential equation based on the following procedure:

- Create a Lefschetz complex X which discretizes the region of interest in phase space. This can be a triangulation or a more general complex. Moreover, any a priori knowledge about the expected dynamics could provide additional guidance that can be exploited for the discretization.
- For every cell $\sigma \in X$ with $\dim \sigma < \dim X$, use the vector field f to find all higher-dimensional cells which can potentially be reached directly from σ. Combine this collection of cells with σ in a set $D_\sigma \subset X$.
- Let $\mathcal{V}$ denote the minimal multivector field guaranteed by Theorem 2.8.1 for the collection $\mathcal{D} = \{D_\sigma \mid \dim \sigma < \dim X\}$.

The resulting multivector field can then be analyzed using the Conley approach based on connection matrices described in the last few sections. Notice that of course we could have $\mathcal{V} = \{X\}$, and in this case the generated multivector field is trivial and would yield no information about the underlying dynamical system φ. Yet, as the following two examples show, usually one does obtain useful insights. In fact, in many situations, these can be extended to computer-assisted proofs for certain dynamical behavior of φ.

Example 2.8.2 (Multivector Field Analysis of Classical Gradient Flows) As our first example, we consider the planar ordinary differential equation

$$\begin{aligned} \dot{x}_1 &= x_1\left(1 - x_1^2 - 3x_2^2\right), \\ \dot{x}_2 &= x_2\left(1 - 3x_1^2 - x_2^2\right). \end{aligned} \tag{2.10}$$

This system is a gradient system, since the right-hand side equals the negative gradient of the potential function

$$V(x_1, x_2) = \frac{x_1^4 + 6x_1^2x_2^2 + x_2^4}{4} - \frac{x_1^2 + x_2^2}{2}.$$

One can show that this nonlinear differential equation has exactly nine equilibrium solutions, which can be grouped as follows:

- The origin is an unstable equilibrium $\mathbf{u}$ with index 2, i.e., it has a two-dimensional unstable manifold.
- The four points $(\pm 1/2, \pm 1/2)$ are all unstable saddle equilibria with Morse index 1, i.e., with one-dimensional unstable manifolds. We denote the saddle in the k-th quadrant by $\mathbf{s}_k$ for $k = 1, \ldots, 4$.
- Finally, there are four asymptotically stable equilibria at the points $(\pm 1, 0)$ and $(0, \pm 1)$. Starting with the stationary state on the positive x_1-axis, and moving in counterclockwise direction, we denote these by $\mathbf{a}_k$ for $k = 1, \ldots, 4$.

A sketch of the phase portrait of this system is shown in Fig. 2.8. The eight nontrivial equilibrium points lie along the corners and edge centers of the diamond given by

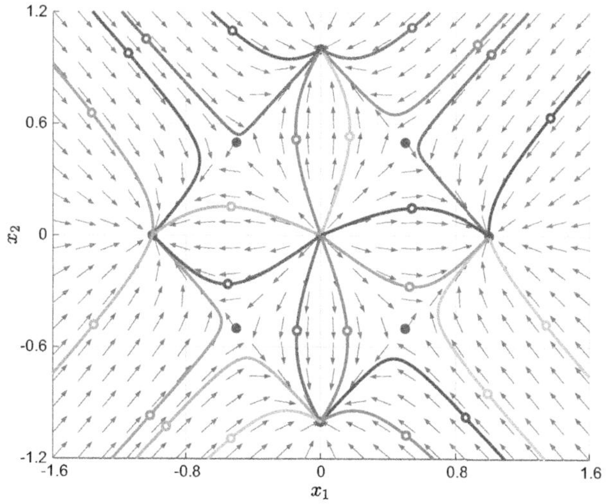

Fig. 2.8 *Multivector field analysis of classical flows.* Sketch of the phase portrait of the planar flow defined in (2.10) in Example 2.8.2. The system is a gradient ordinary differential equation with one unstable equilibrium at the origin, as well as eight further equilibria along the sides of the diamond given by the equation $|x_1| + |x_2| = 1$

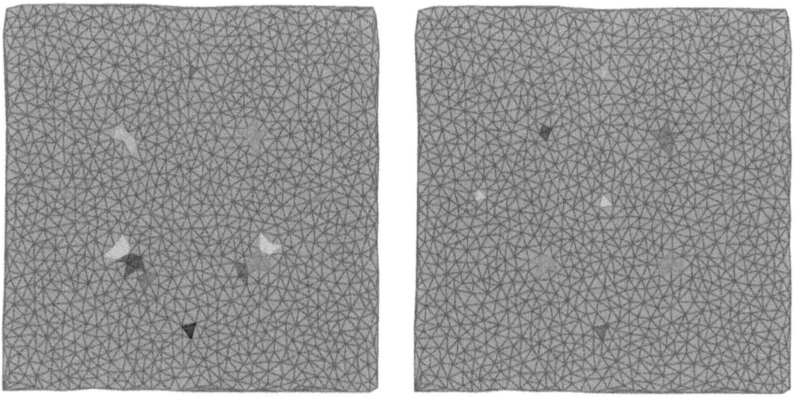

Fig. 2.9 *Multivector field analysis of classical flows.* For the system (2.10) in Example 2.8.2, and the depicted Delaunay triangulation, the left image shows all multivectors of size at least 7 in the multivector field $\mathcal{V}$ guaranteed by Theorem 2.8.1. The image on the right depicts the finest Morse decomposition of $\mathcal{V}$, which consists of nine Morse sets

the equation $|x_1| + |x_2| = 1$. The image also shows the underlying right-hand side vector field, as well as a few sample orbits.

In order to analyze this system using multivector fields, we make use of the Julia [7] package `ConleyDynamics.jl` [49], which allows one to easily create a random Delaunay triangulation X of part of the phase space. Specifically, we consider the square region $U = (-3/2, 3/2)^2 \subset \mathbb{R}^2$ and the triangulation shown in the two panels of Fig. 2.9. Based on this discretization and the vector field defined in (2.10), one can then determine a multivector field $\mathcal{V}$ using the above-described procedure. We would like to point out that for any non-transverse edge in the triangulation the two adjoining triangles have to be part of the same multivector. In other words, the application of Theorem 2.8.1 can certainly lead to large multivectors. As it turns out, however, this is usually not the case. In our example, the multivector field $\mathcal{V}$ consists of 2255 multivectors, of which 2167 are regular with precisely one triangle, that is, have size 2 (a triangle and an edge) or 4 (a triangle, two edges, and one vertex). There are only 12 multivectors of size at least 7, and they are all shown in different colors in the left panel of Fig. 2.9. In fact, the largest multivector in $\mathcal{V}$ consists of merely 21 simplices, only seven of which are triangles.

If we now compute the Morse decomposition and associated connection matrix for the multivector field $\mathcal{V}$ on the simplicial complex X over the field $\mathbb{Z}_2$, then we obtain precisely nine Morse sets, which are illustrated in the right panel of Fig. 2.9. Notice that each Morse set contains exactly one of the equilibrium solutions shown in Fig. 2.8. We would like to point out that each of the four asymptotically stable ones is represented by a multivector of size seven—which is the combinatorial closure of the triangle containing $\mathbf{a}_k$. Moreover, the connection matrix for this combinatorial Morse decomposition is given by

	$\mathbf{s}_1$	$\mathbf{s}_2$	$\mathbf{s}_3$	$\mathbf{s}_4$	$\mathbf{u}$
$\mathbf{a}_1$	1	0	0	1	
$\mathbf{a}_2$	1	1	0	0	
$\mathbf{a}_3$	0	1	1	0	
$\mathbf{a}_4$	0	0	1	1	
$\mathbf{s}_1$					1
$\mathbf{s}_2$					1
$\mathbf{s}_3$					1
$\mathbf{s}_4$					1

Note that every heteroclinic orbit between $\mathbf{u}$ and the $\mathbf{s}_k$ as well as between the $\mathbf{s}_k$ and the $\mathbf{a}_k$ that is indicated in Fig. 2.8 is in fact represented by a nonzero entry in the connection matrix.

Yet, even more can be gleaned from the multivector field. To explain this, let P denote the poset associated with the Morse decomposition of the multivector field $\mathcal{V}$, and let $I \subset P$ denote an arbitrary convex subset. Then the *Morse interval* M_I is defined as the collection of all cells that lie on paths which originate and end in the union $\bigcup_{p \in I} M_p$. A special case of this notion for down sets I has already been introduced in the lead-up to Example 2.2.5. The Julia package `ConleyDynamics.jl` can compute Morse intervals, and a few of them are illustrated in Fig. 2.10. While the left panel shows M_I for the down set $I = \{\mathbf{s}_k, \mathbf{a}_k \mid k = 1, \ldots, 4\}$, the panel on the right contains, in addition to the Morse sets, also the Morse intervals for the down set $I = \{\mathbf{s}_3, \mathbf{a}_4\}$ and for the convex subset $I = \{\mathbf{u}, \mathbf{s}_1, \mathbf{s}_4\}$. These Morse intervals are rough enclosures for some of the

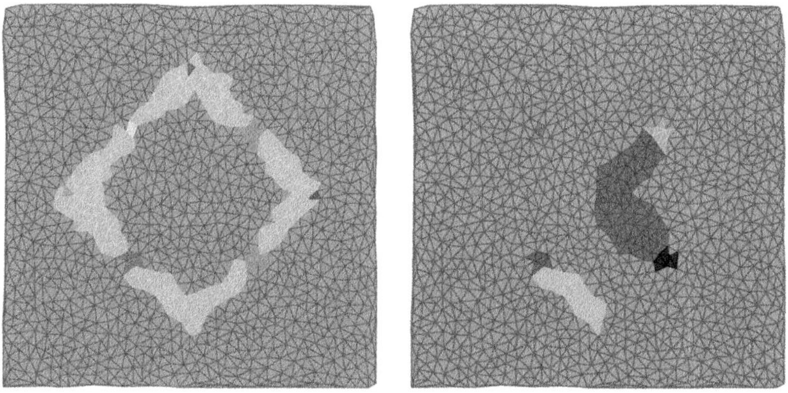

Fig. 2.10 *Multivector field analysis of classical flows*. For the multivector field $\mathcal{V}$ underlying Fig. 2.9, the two images illustrate a variety of Morse intervals M_I. The left panel is for the down set $I = \{\mathbf{s}_k, \mathbf{a}_k \mid k = 1, \ldots, 4\}$, while the panel on the right illustrates the Morse sets, as well as the Morse intervals for the down set $I = \{\mathbf{s}_3, \mathbf{a}_4\}$ and the convex set $I = \{\mathbf{u}, \mathbf{s}_1, \mathbf{s}_4\}$

above-mentioned heteroclinic orbits. In fact, in combination with interval arithmetic computations, they can be used to prove the existence of these connections in the original system (2.10). The details for this can be found in [48]. ◊

Example 2.8.3 (Multivector Field Analysis of Classical Recurrent Flows) As our second and final example, we consider the planar system of ordinary differential equations given by

$$\begin{aligned} \dot{x}_1 &= +x_2 - x_1 \left(x_1^2 + x_2^2 - 4\right)\left(x_1^2 + x_2^2 - 1\right), \\ \dot{x}_2 &= -x_1 - x_2 \left(x_1^2 + x_2^2 - 4\right)\left(x_1^2 + x_2^2 - 1\right). \end{aligned} \tag{2.11}$$

This system is no longer of gradient type, and therefore it allows for the possibility of periodic solutions. In fact, one can show that the resulting flow has an asymptotically stable equilibrium at the origin, as well as both an unstable and a stable periodic orbit surrounding it. The system's phase portrait is illustrated in Fig. 2.11.

Analogous to the previous example, we use `ConleyDynamics.jl` to determine a multivector field $\mathcal{V}$ as guaranteed by Theorem 2.8.1 and our outlined procedure based on dynamical transitions. For the random Delaunay triangulation shown in Fig. 2.12, this results in a multivector field $\mathcal{V}$ which has a total of 2170 multivectors. Of these, there are 2002 regular multivectors with exactly one triangle, almost equally split between size 2 and 4. As in the previous example, large multivectors are rare. The largest one consists of only 14 cells, and in fact there are merely 30 multivectors of size at least 7, as shown in the left panel of Fig. 2.12. With one exception, all multivectors are regular multivectors.

The finest Morse decomposition for the multivector field $\mathcal{V}$ is shown in the right panel of Fig. 2.12, and it contains the expected three Morse sets. Their Conley

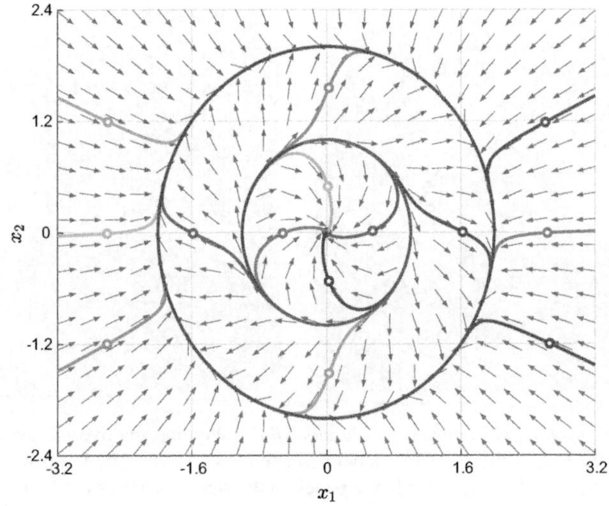

Fig. 2.11 *Multivector field analysis of classical flows.* Sketch of the phase portrait of the planar flow defined in (2.11) in Example 2.8.3. The system has two circular periodic orbits of radii 1 and 2 which surround an asymptotically stable equilibrium at the origin

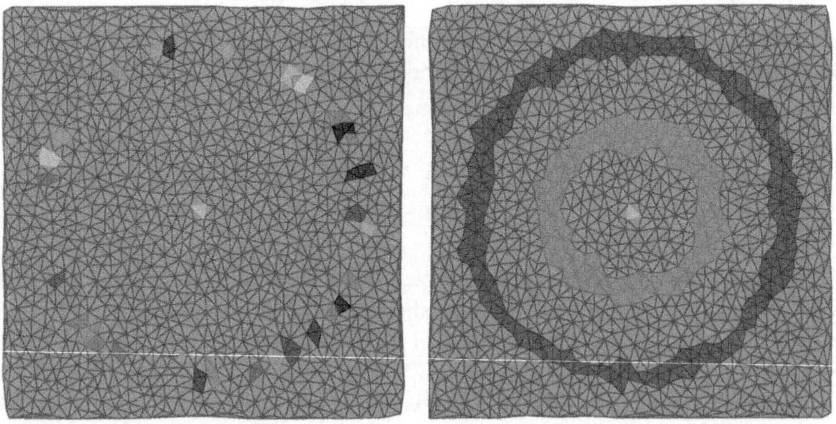

Fig. 2.12 *Multivector field analysis of classical flows.* For the system (2.11) in Example 2.8.3 and the depicted Delaunay triangulation, the left image shows all 30 multivectors of size at least 7 in the multivector field $\mathcal{V}$ guaranteed by Theorem 2.8.1. The image on the right depicts the finest Morse decomposition of $\mathcal{V}$, which consists of three Morse sets

polynomials are 1 for the equilibrium $\mathbf{a}$ at the origin, $1 + t$ for the stable periodic orbit $\mathbf{s}$ with radius 2, and $t + t^2$ for the unstable periodic orbit $\mathbf{u}$ with radius 1. Finally, the associated connection matrix is given by

	$\mathbf{a}^0$	$\mathbf{s}^0$	$\mathbf{s}^1$	$\mathbf{u}^1$	$\mathbf{u}^2$
$\mathbf{a}^0$				1	0
$\mathbf{s}^0$				1	0
$\mathbf{s}^1$				0	1

.

In this formula, the superscripts indicate the dimension of the corresponding Conley index generator. Clearly, the matrix indicates the connecting orbits between the unstable periodic orbit $\mathbf{u}$ and both the periodic orbit $\mathbf{s}$ and the equilibrium $\mathbf{a}$. Furthermore, one can use the techniques of [37] to rigorously establish the existence of the two periodic orbits in the two large Morse sets. For details, we refer the reader again to [48]. ◊

These two examples are only meant as a brief illustration of the immense potential that is provided by the use of combinatorial multivector fields for the analysis of classical dynamics. For further examples, we refer the reader to [37, 48, 49]. We would like to point out, however, that these simple examples produced useful insight even for fairly coarse triangulations of the underlying phase space region. One can expect that more sophisticated discretization techniques should lead to much tighter enclosures of the dynamics of interest.

2.9 Connection Matrices in Classical Flows

The discussion in the previous section raises the question of what is the relation between the connection matrices in their classical setting of flows and the connection matrices in the combinatorial setting of multivector fields. The answer may be found, in a somewhat hidden way, in the Robbin and Salamon construction of connection matrices for flows [44]. In order to explain this in more detail, consider a smooth dynamical system together with a Morse decomposition $\mathcal{M} := \{M_p\}_{p \in P}$ indexed by a finite poset P. Then in view of [44, Theorem 4.2] there exists a lattice homomorphism $I \mapsto N_I$ which assigns to each down set I of P an attracting neighborhood N_I isolating an attractor A_I. Moreover, within the proof of [44, Theorem 7.3], it is shown, based on a method of Cairns [10], that one can triangulate the phase space and choose each N_I in such a way that it is a subcomplex of this triangulation. The triangulation, considered as the family of simplices of the simplicial complex, is a Lefschetz complex. We now denote this Lefschetz complex by K, and the subcomplex presentation of N_I by $K_I \subset K$. Then the family $\mathcal{K} := \{ K_I \mid I \in \text{Down}(P) \}$ is a lattice, the mapping $I \mapsto K_I$ is a lattice isomorphism, and, as was already explained in Sect. 2.3 (see also [44, Proposition 4.1 and Theorem 4.2]), the collection $\mathcal{V} := \text{AP}(\mathcal{K})$ given by (2.5) forms an acyclic partition of K. Each set in $\mathcal{V}$ is of the form $V_p := K_{p^{\leq}} \setminus K_{p^{<}}$, where $p^{\leq} := \{q \in P \mid q \leq p\}$ and $p^{<} := p^{\leq} \setminus \{p\}$ for some $p \in P$. This is due to the fact that every join-irreducible $I \in \mathrm{Down}(P)$ is of the form $p^{\leq}$ for some $p \in P$ and then $I^{\star} = p^{<}$. It follows that each V_p is locally closed as a difference of two closed subsets of K. Therefore, the collection $\mathcal{V}$ is in fact a combinatorial multivector field on K. Since $\mathcal{V}$ forms an acyclic partition, it is also a Morse decomposition of $\mathcal{V}$, and its connection matrix is the connection matrix of the Morse decomposition $\mathcal{M}$. This connection matrix coincides with the one defined in [44] for flows, because both are defined via the filtered chain complex associated with the lattice $\mathcal{K}$.

Summarizing, connection matrices for combinatorial multivector fields naturally arise from the construction of connection matrices for flows by Robbin & Salamon. The significance of [44] lies in ensuring that the connection matrices for flows are well-defined. However, it is far from obvious how one can turn their construction into an algorithm. Therefore, for concrete cases, one still needs to find a triangulated lattice of attracting neighborhoods in order to compute a connection matrix of a flow. The connection matrix algorithms introduced in [13, 46] assume such a lattice as given. Yet, to the best of our knowledge, no algorithm is known for the construction of such a lattice from a vector field. The following example indicates some of the challenges that can be encountered and also suggests that the theory of connection matrices for combinatorial multivector fields may turn out to be helpful in computing connection matrices for flows.

Example 2.9.1 (A Flow Leading to a Multivector Field Without Lattice of Attractors) Consider a planar vector field $h : \mathbb{R}^2 \to \mathbb{R}^2$ given by

$$h(x, y) := \varphi(x, y) f(x, y) + (1 - \varphi(x, y)) g(x, y), \tag{2.12}$$

where

$$f(x, y) := ((x - 5 \operatorname{sgn} x) \operatorname{sgn} y, \operatorname{sgn} y(5 \operatorname{sgn} y - y)), \tag{2.13}$$

$$g(x, y) := \left(-\frac{2xy}{y^2 + 1} \min\left(1, \frac{y^2 + 1}{|x|}\right), -1 \right), \tag{2.14}$$

$$\varphi(x, y) := \min\left(1, (\max(1, \min(|x|, |y|)) - 1)/2\right). \tag{2.15}$$

Since h is Lipschitz continuous, it generates a flow on $\mathbb{R}^2$. This flow has a Morse decomposition consisting of four stationary points, one per each quadrant of the plane. The phase portrait of this flow, presented in the left panel of Fig. 2.13, is similar to the phase portrait of the multiflow discussed in Example 2.2.5. In particular, both the multiflow in the right panel of Fig. 2.3 and the flow in the left image of Fig. 2.13 have a minimal Morse decomposition which consists of four saddle points. However, the Morse decomposition of the flow, unlike the Morse decomposition of the multiflow, does admit a lattice of attracting neighborhoods, as follows from [44, Theorem 4.2]. Nevertheless, constructing it may not be

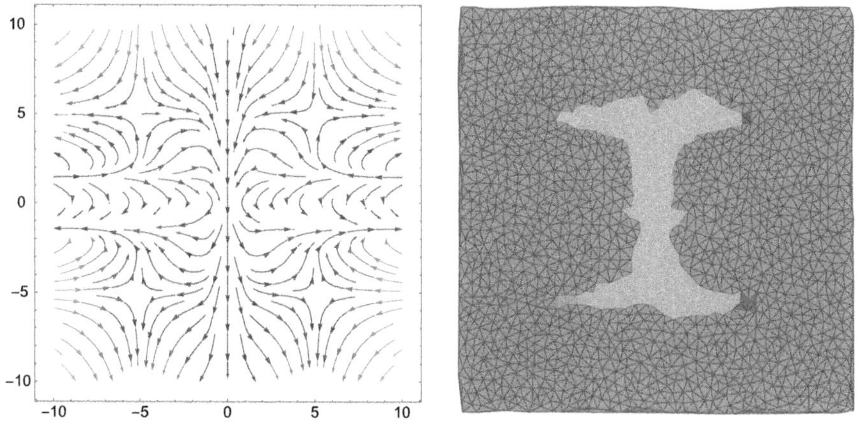

Fig. 2.13 *A flow leading to a multivector field without lattice of attractors.* The image on the left contains the phase portrait of the flow induced by the vector field (2.12)–(2.15). The panel on the right depicts a Morse decomposition of the maximal invariant set in a multivector field which was constructed from this flow, based on a random Delaunay triangulation, by means of Theorem 2.8.1

easy. Notably, an attempt to construct a multivector field on a random Delaunay triangulation leads to a Morse decomposition consisting of four Morse sets, each with its Conley index of a saddle, but without lattice of attractors. The connection matrix is zero for both the flow and the combinatorial multivector field. In the latter case, its computation is straightforward, but it requires a temporary expansion of the poset similarly to the discussion in Example 2.3.2, which was continued in Example 2.5.1. ◊

The example indicates that the simplified pipeline presented in this book, which takes any family of attracting neighborhoods and extends it to a lattice, may be useful also for computing connection matrices in the classical setting of flows. In practice, it is even simpler to discretize the phase space, apply Theorem 2.8.1 to construct a combinatorial multivector field guaranteeing transversal intersections of cells of codimension one, choose a Morse decomposition of interest for this multivector field, compute its connection matrix via the associated acyclic partition, and then carry it over to the original flow by applying [37, Theorem 5.28]. For more details we refer the reader to the discussion in [48].

Chapter 3
Algebraic and Topological Background Material

In this chapter, we collect background material that is necessary for the more technical parts of the book. This includes both definitions and fundamental results. We begin with a discussion of relations and partial orders, before moving on to topological spaces. The following two sections are more algebraic in nature, as they address modules, chain complexes, and homology. Finally, the last section is concerned with a detailed survey of Lefschetz complexes.

3.1 Sets, Maps, Relations, and Partial Orders

We denote the sets of reals, integers, positive integers, and nonnegative integers by $\mathbb{R}$, $\mathbb{Z}$, $\mathbb{N}$, and $\mathbb{N}_0$, respectively. Given a set X, we write $\operatorname{card} X$ for the number of elements of X, and we denote the family of all subsets of X by $\mathcal{P}(X)$.

A *P-indexed family* $(X_p)_{p\in P}$ of subsets of an arbitrary set X is a mapping of the form $P \ni p \mapsto X_p \in \mathcal{P}(X)$. Note that every family $\mathcal{E} \subset \mathcal{P}(X)$ may be considered as an $\mathcal{E}$-indexed family via the identity map $\mathcal{E} \ni E \mapsto E \in \mathcal{P}(X)$. In this case, we say that the family $\mathcal{E}$ is a *self-indexed* family. A *partition* of a set X is an indexed family $\mathcal{E} = (E_p)_{p\in P} \subset \mathcal{P}(X)$ of mutually disjoint subsets of X which satisfies the identity $\bigcup_{p\in P} E_p = X$. Given a partition $(E_p)_{p\in P}$ and a subset $I \subset P$, we will use the compact notation $|I| := \bigcup_{p\in I} E_p$.

By a *K-gradation* of a set X, we mean a K-indexed family $(X_k)_{k\in K}$ such that $k \mapsto X_k$ is injective and $(X_k)_{k\in K}$ is a partition of X. Note, in particular, that every surjective map $f : X \to K$ induces the K-gradation $(f^{-1}(k))_{k\in K}$ of X. We refer to this K-gradation as the *f-gradation of* X. Given a subset $A \subset X$ and a gradation $(X_k)_{k\in K}$ of X, the *induced gradation* is the gradation $(A \cap X_k)_{k\in K_A}$ of A where $K_A := \{k \in K \mid A \cap X_k \neq \varnothing\}$.

We write $f : X \nrightarrow Y$ for a *partial map* from X to Y, i.e., a map defined on a subset $\operatorname{dom} f \subset X$, called the *domain* of f, and such that the set of values

M. Mrozek, T. Wanner, *Connection Matrices in Combinatorial Topological Dynamics*, SpringerBriefs in Mathematics,
https://doi.org/10.1007/978-3-031-87600-4_3

of f, denoted by $\operatorname{im} f$, is contained in Y. Partial maps are composed in the obvious way: If $f : X \nrightarrow Y$ and $g : Y \nrightarrow Z$ are partial maps, then the domain of their composition $g \circ f : X \nrightarrow Z$ is given by $\operatorname{dom}(g \circ f) = f^{-1}(\operatorname{dom} g)$, and on this domain, we have $(g \circ f)(x) = g(f(x))$.

In the following, we work with the *category of finite sets with a distinguished subset*, which is denoted by DSET and defined as follows. The objects of DSET are pairs $(X, X_\star)$, where X is a finite set and $X_\star \subset X$ is a distinguished subset. The morphisms from $(X, X_\star)$ to $(Y, Y_\star)$ in DSET are *subset preserving partial maps*, that is, partial maps $f : X \nrightarrow Y$ such that the inclusions $X_\star \subset \operatorname{dom} f$ and $f(X_\star) \subset Y_\star$ are satisfied. One easily verifies that DSET with the composition of morphisms defined as the composition of partial maps and the identity morphism defined as the identity map is indeed a category.

Given a set X and a binary relation $R \subset X \times X$, we write xRy meaning $(x, y) \in R$. The *inverse* of R, denoted R^{-1}, is the relation $R^{-1} := \{(y, x) \mid xRy\}$. By the *transitive closure* of R, we mean the relation $\bar{R} \subset X \times X$ given by $x\bar{R}y$ if there exists a sequence $x = x_0, x_1, \ldots, x_n = y$ such that $n \geq 0$ and $x_{i-1}Rx_i$ for $i = 1, 2, \ldots, n$.

A *multivalued map* $F : X \multimap Y$ is a map $F : X \to \mathcal{P}(Y)$. For $A \subset X$, we define the *image of* A by $F(A) := \bigcup\{F(x) \mid x \in A\}$, and for $B \subset Y$, we define the *preimage of* B by $F^{-1}(B) := \{x \in X \mid F(x) \cap B \neq \varnothing\}$.

Given a relation R, we associate with it a multivalued map $F_R : X \multimap X$ via the definition $F_R(x) := R(x)$, where $R(x) := \{y \in X \mid xRy\}$ is the *image of* $x \in X$ *in* R. Obviously, $R \mapsto F_R$ is a one-to-one correspondence between binary relations in X and multivalued maps from X to X. Often, it will be convenient to interpret the relation R as a directed graph whose set of vertices is X and a directed arrow goes from x to y whenever xRy. The three concepts relation, multivalued map, and directed graph are equivalent on the formal level, and the distinction is used only to emphasize different points of view. However, in this book, it will be convenient to use all of these concepts interchangeably.

Recall that a relation R is a *partial order* if it is reflexive, antisymmetric, and transitive. If not stated otherwise, we denote a partial order by $\leq$ and its inverse by $\geq$. We also write $<$ and $>$ for the associated strict partial orders, that is, the relations $\leq$ and $\geq$ excluding equality. We say that R is *acyclic* if there are no sequences $x_0, x_1, \ldots, x_n = x_0$ such that $n > 0$ and $x_{i-1}Rx_i$ for $i = 1, 2, \ldots, n$. The following straightforward proposition characterizes acyclic relations.

Proposition 3.1.1 *A relation R is acyclic if and only if its transitive closure is a partial order.* □

We recall that a *poset* or *partially ordered set* is a pair $\mathbb{P} = (P, \leq)$ where P is a set and $\leq$ is a partial order in P. Whenever the partial order is clear from context, we refer to P as the poset.

Given a partially ordered set P, we say that $q \in P$ *covers* $p \in P$ if $p < q$, and there is no $r \in P$ such that $p < r < q$. We say that p is a *predecessor* of q if q covers p. Recall that $A \subset P$ is *convex* if $x, z \in A$ and $x \leq y \leq z$ imply $y \in A$. Clearly, the intersection of a family of convex sets is a convex set. This lets us define the *convex hull* of $J \subset P$, denoted $\operatorname{conv}_P(J)$, as the intersection of the family of all convex sets in P containing J. It is straightforward to observe that

$$\operatorname{conv}_P(J) = \{ p \in P \mid \exists p_-, p_+ \in J \text{ with } p_- \le p \le p_+ \}. \tag{3.1}$$

A set $A \subset P$ is a *down set* or a *lower set* if the two conditions $z \in A$ and $y \le z$ imply the inclusion $y \in A$. Dually, the subset $A \subset P$ is an *upper set* if $z \in A$ and $z \le y$ yield $y \in A$.

We denote the family of all down sets in P by $\operatorname{Down}(P)$. For $A \subset P$, we further write $A^{\le} := \{ x \in P \mid \exists_{a \in A}\, x \le a \}$ and $A^{<} := A^{\le} \setminus A$.

Proposition 3.1.2 *For any subset $I \subset P$, the set $I^{\le}$ is a down set. In addition, if I is convex, then $I^{<}$ is a down set as well.*

Proof The verification that $I^{\le}$ is a down set is straightforward. To see that $I^{<}$ is a down set, take an $x \in I^{<}$. Then we have $x \notin I$ and $x < z$ for some $z \in I$. Let $y \le x$. Then $y \in I^{\le}$. Since I is convex, we cannot have $y \in I$. It follows that $y \in I^{<}$. □

Let P and P' be posets. We say that a partial map $f : P \nrightarrow P'$ is *order preserving* if the conditions $x, y \in \operatorname{dom} f$ and $x \le y$ imply $f(x) \le f(y)$. We say that f is an *order isomorphism* if f is an order preserving bijection such that f^{-1} is also order preserving.

We define the category DPSET of posets with a distinguished subset as follows. Its objects are pairs $(P, P_\star)$ where P is a finite poset and $P_\star \subset P$ is a distinguished subset. A morphism from $(P, P_\star)$ to $(P', P'_\star)$ in DPSET is an order preserving partial map $f : P \nrightarrow P'$ such that $P_\star \subset \operatorname{dom} f$ and $f(P_\star) \subset P'_\star$. We call such a morphism f *strict* if $\operatorname{dom} f = P_*$. One easily verifies that DPSET with the composition of morphisms defined as the composition of partial maps and the identity morphism defined as the identity map is indeed a category. Moreover, since every set may be considered as a poset partially ordered by the identity and, clearly, every partial map preserves identities, we may consider DSET as a subcategory of DPSET. In this case, every partial map is automatically order preserving. To simplify notation, in the sequel, we denote objects of DSET and DPSET with a single capital letter and the distinguished subset by the same letter with subscript $\star$.

3.2 Topological Spaces

We refer the reader to [17, 40] for our terminology concerning general topological spaces and to [39] for the terminology concerning algebraic topology, homological algebra, and category theory. Here we recall only the basic concepts and emphasize differences in notation.

A *topology* on a set X is a family $\mathcal{T}$ of subsets of X which is closed under finite intersections and arbitrary unions, and which satisfies $\varnothing, X \in \mathcal{T}$. A *topological space* is a pair $(X, \mathcal{T})$ where $\mathcal{T}$ is a topology on X. We often refer to X as a topological space assuming that the topology $\mathcal{T}$ on X is clear from context. The sets in $\mathcal{T}$ are called *open*. The *interior* of A, denoted $\operatorname{int} A$, is the union of all open subsets of A. A subset $A \subseteq X$ is *closed* if $X \setminus A$ is open. The *closure* of A, denoted $\operatorname{cl} A$, is the intersection of all closed supersets of A.

Every subset $Y \subset X$ of a topological space X is a topological space itself with the *induced topology* given as the family $\{U \cap Y \mid U \text{ open in } X\}$. Given two topological spaces X and X', we say that $f : X \to X'$ is *continuous* if $f^{-1}(U)$ is open in X for every U open in X'.

If A is a subset of a topological space X, we denote the difference $\operatorname{cl} A \setminus A$ by $\operatorname{mo} A$, and we call it the *mouth* of A. The concept of mouth is frequently used in this book together with the concept of a locally closed set recalled in the following proposition and definition.

Proposition and Definition 3.2.1 (See Problem 2.7.1 in [17]) *Assume that A is a subset of a topological space X. Then the following conditions are equivalent:*

(i) Each $x \in A$ admits a neighborhood U in X such that $A \cap U$ is closed in U.
(ii) The mouth of A is closed in X, where $\operatorname{mo} A = \operatorname{cl} A \setminus A$.
(iii) A is a difference of two closed subsets of X.
(iv) A is an intersection of an open set in X and a closed set in X.

If A satisfies any of the equivalent conditions (i)–(iv), then A is called locally closed.

Proof It is obvious that (ii) implies (iii), (iii) implies (iv), and that (iv) implies (i). Thus, it suffices to prove that if (ii) fails, then so does (i). Hence, we now assume that the mouth $\operatorname{mo} A = \operatorname{cl} A \setminus A$ is not a closed set in X. Then it is clear that there has to exist a point $x \in \operatorname{cl}(\operatorname{cl} A \setminus A) \setminus (\operatorname{cl} A \setminus A)$. It follows that $x \in A$. Now let U be an arbitrary neighborhood of x in X. We will establish that $A \cap U$ is not closed in U. Since $x \in \operatorname{cl}(\operatorname{cl} A \setminus A)$, we can find a point $y \in (\operatorname{cl} A \setminus A) \cap U$. Let V be an arbitrary open neighborhood of y in X. Then also the intersection $V \cap U$ is a neighborhood of y in X. Since $y \in \operatorname{cl} A \setminus A \subset \operatorname{cl} A$, we obtain $V \cap U \cap A \neq \varnothing$. Since all open neighborhoods of y in U are of the form $V \cap U$ where V is open in X, we see that y is in the closure of $U \cap A$ in U. However, $y \notin A$, hence, $y \notin U \cap A$. It follows that the intersection $U \cap A$ is not closed in U. This holds for every neighborhood U of x in X. Therefore, (i) fails. □

The topology $\mathcal{T}$ is T_2 or *Hausdorff* if for any two different points $x, y \in X$ there exist disjoint sets $U, V \in \mathcal{T}$ such that $x \in U$ and $y \in V$. It is T_0 or *Kolmogorov* if for any two different points $x, y \in X$ there exists a $U \in \mathcal{T}$ such that $U \cap \{x, y\}$ is a singleton.

Finally, we say that a topological space X is a *finite topological space* if the underlying set X is finite. Finite topological spaces differ from general topological spaces, because the only Hausdorff topology on a finite topological space X is the discrete topology consisting of all subsets of X. Moreover, the celebrated Alexandrov Theorem [1] states that every finite, T_0 topological space can equivalently be considered as a poset, if we set $x \leq_{\mathcal{T}} y$ whenever $x \in \operatorname{cl} y$. The results of this book apply only to a very special, but crucial for applications, class of finite topological spaces, namely Lefschetz complexes (see Sect. 3.5). However, the view through the lens of the nonstandard features of finite topological spaces facilitates a better understanding of the peculiarities of combinatorial topological dynamics. We refer the interested reader to [2–4, 30] for more information.

3.3 Modules and Their Gradations

For the terminology used in the book concerning modules, we refer the reader to [28]. Here, we briefly recall that a *module* over a ring R is a triple $(M, +, \cdot)$ such that $(M, +)$ is an abelian group and $\cdot : R \times M \to M$ is a scalar multiplication which is distributive with respect to the additions in M and R, and which satisfies both $a \cdot (b \cdot x) = (ab) \cdot x$ and $1_R \cdot x = x$ for $x \in M$ and $a, b \in R$. We shorten the notation $a \cdot x$ to ax. We often say that M is a module assuming that the operations $+$ and $\cdot$ are clear from context. Recall that a *submodule* of M is a subset $N \subset M$ such that N with the operations $+$ and $\cdot$ restricted to N is itself a module. The associated *quotient module* is denoted M/N. For submodules $N_1, N_2, \ldots, N_k$ of a given module M, their *algebraic sum*

$$N_1 + N_2 + \cdots N_k := \{ x_1 + x_2 + \ldots + x_k \in M \mid x_i \in N_i \}$$

is easily seen to be a submodule of M. We call it a *direct sum* and denote it by the symbol $N_1 \oplus N_2 \oplus \ldots \oplus N_k$, if for every element $x \in N_1 + N_2 + \ldots + N_k$ the representation $x = x_1 + x_2 + \ldots + x_k$ with $x_i \in N_i$ is unique.

The following straightforward result will be useful for establishing the existence of connection matrices later on.

Proposition 3.3.1 *Assume that A and B are submodules of an arbitrary module X which satisfy $X = A + B$. If B' is a submodule of B such that $B = (A \cap B) \oplus B'$, then we have $X = A \oplus B'$.* □

A subset $Z \subset M$ *generates* the module M if for every element $x \in M$ there exist elements $x_1, x_2, \ldots, x_n \in Z$ and scalars $a_1, a_2, \ldots, a_n \in R$ such that $x = \sum_{i=1}^{n} a_i x_i$. A subset $Z \subset M$ is *linearly independent* if for any choice of $x_1, x_2, \ldots, x_n \in Z$ and scalars $a_1, a_2, \ldots, a_n \in R$ the equality $\sum_{i=1}^{n} a_i x_i = 0$ implies $a_1 = \ldots = a_n = 0$. A linearly independent subset $B \subset M$ which generates M is called a *basis* of M. A module may not have a basis. If it does have a basis, the module is called *free*. We note that $\varnothing$ is the unique basis of the trivial module $\{0\}$.

Assume that M is a free module and that $B \subset M$ is a fixed basis. Then we have the associated bilinear map $\langle \cdot, \cdot \rangle_B : M \times M \to R$, called *scalar product*, and defined on basis elements $b, b' \in B$ by letting $\langle b, b' \rangle_B := 0$ for $b \neq b'$, as well as $\langle b, b' \rangle_B := 1$ for $b = b'$. For an arbitrary element $x \in M$, we define its *support* with respect to the basis B as $|x|_B := \{ b \in B \mid \langle x, b \rangle_B \neq 0 \}$. One easily verifies that for all $x, y \in M$

$$|x + y|_B \subset |x|_B \cup |y|_B. \tag{3.2}$$

Assume now that X is any finite set. Then the collection $R\langle X\rangle := \{\, f : X \to R \,\}$ of all R-valued functions forms a module with respect to pointwise addition and scalar multiplication. We note that $R\langle\varnothing\rangle$ is the zero module. For every $x \in X$ we have a function $\bar{x} : X \to R$ which sends x to 1 and any other element of X to 0. One easily verifies that $\bar{X} := \{\, \bar{x} \mid x \in X \,\}$ is a basis of $R\langle X\rangle$. Therefore, $R\langle X\rangle$ is a free module, and we call $R\langle X\rangle$ the *free module spanned by* X. In the sequel, we identify $\bar{x}$ with x.

Let M and M' be two modules over R. A map $h : M \to M'$ is called a *module homomorphism* if $h(a_1x_1 + a_2x_2) = a_1h(x_1) + a_2h(x_2)$ for $x_1, x_2 \in M$ and $a_1, a_2 \in R$. The *kernel* of h, denoted by $\ker h$, is the submodule $\{\, x \in M \mid h(x) = 0 \,\}$, while the *image* of h, denoted by $\operatorname{im} h$, is the submodule $\{\, y \in M' \mid \exists_{x \in M} h(x) = y \,\}$.

We now turn our attention to gradations of modules. Assume that R is a fixed ring and M is a module over R. Let K be an arbitrary index set. Then a K-*gradation* of M is a collection of submodules $(M_k)_{k\in K}$ of M indexed by K such that M is the direct sum of the submodules M_k, i.e., we have the identity $M = \bigoplus_{k\in K} M_k$. This definition requires that $K \neq \varnothing$, but it is convenient to assume that the direct sum over the empty set is always the zero module. In view of this, the only module admitting the $\varnothing$-gradation is the zero module. By a K-*graded module* over R, we mean a module M over R together with a fixed, implicitly given, K-gradation. Note that the zero module admits not only the $\varnothing$-gradation, but also a unique K-gradation for each non-empty set K. Thus, for each set K, the zero module is a K-graded module. We denote it by 0_K.

If B is a basis of a free module M and $\mathcal{A}$ is a partition of B, then M is $\mathcal{A}$-graded with the gradation

$$M = \bigoplus_{A\in\mathcal{A}} R \cdot A,$$

where the expression on the right-hand side is defined as

$$R \cdot A = \left\{ \sum_{i=1}^{n} r_i a_i \;\middle|\; r_i \in R,\ a_i \in A,\ i = 1, \ldots, n,\ n \in \mathbb{N} \right\}.$$

Note that in the case that A is finite, one can identify $R \cdot A$ with $R\langle A\rangle$. As a special case, consider the situation when the partition of B consists only of singletons. Then the gradation becomes

$$M = \bigoplus_{b\in B} R \cdot b.$$

We refer to this gradation as the B-*basis gradation* of M.

We say that a submodule M' of M is a *K-graded submodule* if M' is also K-graded with the decomposition

$$M' = \bigoplus_{k \in K} M'_k$$

and such that $M'_k \subset M_k$. For $I \subset K$, we have an *I-graded submodule*

$$M_I := \bigoplus_{i \in I} M_i. \tag{3.3}$$

The homomorphisms

$$\iota_I : M_I \ni x \mapsto x \in M$$

and

$$\pi_I : M = \bigoplus_{k \in K} M_k \ni \sum_{k \in K} x_k \mapsto \sum_{i \in I} x_i \in M_I$$

are called the associated *canonical inclusion* and *canonical projection*. In the case of a singleton $I = \{ i \}$, we abbreviate the above notation to ι_i and π_i.

Assume K' is another set and M' is a K'-graded module. For a module homomorphism $f : M \to M'$ and subsets $J \subset K$ and $I \subset K'$, we have an *induced homomorphism* $f_{IJ} : M_J \to M'_I$ defined as the composition

$$f_{IJ} := \pi_I \circ f \circ \iota_J,$$

where $\pi_I : M' \to M'_I$ and $\iota_J : M_J \to M$. Again, if $I = \{i\}$ and $J = \{j\}$, we abbreviate the notation to f_{ij}. Then one can easily verify the following proposition.

Proposition and Definition 3.3.2 *Assume K and K' are finite. Then*

$$f = \sum_{i \in K'} \iota_i \sum_{j \in K} f_{ij} \pi_j. \tag{3.4}$$

In particular, the homomorphism f is uniquely determined by its matrix $[f_{ij}]_{i \in K', j \in K}$ of homomorphisms $f_{ij} : M_j \to M'_i$. We refer to this matrix as the (K, K')-matrix of the homomorphism f. □

Notice that in the case when the K- and K'-gradations are basis gradations given by bases $B = \{b_1, b_2, \dots, b_n\}$ in M and $B' = \{b'_1, b'_2, \dots, b'_m\}$ in M', respectively, then the homomorphisms f_{ij} take the form

$$f_{ij} : t \cdot b_j \mapsto t\langle f(b_j), b_i \rangle \cdot b_i,$$

which means that the (K, K')-matrix of f may be identified in this case with the matrix of coefficients $\langle f(b_j), b_i\rangle$. As in the case of classical matrices, one easily verifies the following proposition.

Proposition 3.3.3 *Consider the three modules M, M', and M'', which are graded by K, K', and K'', respectively. Let $f : M \to M'$ and $f' : M' \to M''$ be module homomorphisms. Then the (K, K'')-matrix of f consists of the homomorphisms*

$$(f'f)_{ik} = \sum_{j\in K'} f'_{ij} f_{jk} \tag{3.5}$$

for all $i \in K''$ and $k \in K$. □

Given a $\mathbb{Z}$-graded module M, a homomorphism $f : M \to M$ is of *degree* $k \in \mathbb{Z}$ if $f_{ij} \neq 0$ implies $i - j = k$ for $i, j \in \mathbb{Z}$. If f is of degree k, we briefly write f_j instead of f_{ij} with $i = j + k$.

3.4 Chain Complexes and Their Homology

A *chain complex* is a pair (C, d) consisting of a $\mathbb{Z}$-graded R-module $C = (C_k)_{k\in\mathbb{Z}}$ and a $\mathbb{Z}$-graded homomorphism $d : C \to C$ which satisfies $d^2 = 0$, and which has degree -1, i.e., which satisfies $d(C_k) \subset C_{k-1}$ for all $k \in \mathbb{Z}$. The homomorphism d is called the *boundary homomorphism* of the chain complex (C, d). Let (C', d') be another chain complex. A *chain map* $\varphi : (C, d) \to (C', d')$ is a module homomorphism $\varphi : C \to C'$ of degree zero, satisfying $\varphi d = d'\varphi$. The following proposition is straightforward.

Proposition 3.4.1 *If $\varphi : (C, d) \to (C', d')$ is a chain map between chain complexes and φ is an isomorphism of $\mathbb{Z}$-graded modules, then φ^{-1} is also a chain map.* □

We denote by Cc the category whose objects are chain complexes and whose morphisms are chain maps. One easily verifies that this is indeed a category.

A subset $C' \subset C$ is a *chain subcomplex* of C if C' is a $\mathbb{Z}$-graded submodule of C such that $d(C') \subset C'$. Recall that given a subcomplex $C' \subset C$ we have a well-defined *quotient complex* $(C/C', d')$, where $d' : C/C' \to C/C'$ is the boundary homomorphism induced by d.

We now turn our attention to a fundamental equivalence relation on chain maps. A pair of chain maps $\varphi, \varphi' : (C, d) \to (C', d')$ is called *chain homotopic* if there exists a *chain homotopy* joining φ and φ', i.e., a module homomorphism $S : C \to C'$ of degree $+1$ such that $\varphi' - \varphi = d'S + Sd$. The existence of a chain homotopy between two chain maps is easily seen to be an equivalence relation in the set of chain maps $\text{Cc}((C, d), (C', d'))$. Given a chain map $\varphi \in \text{Cc}((C, d), (C', d'))$, we denote by $[\varphi]$ the equivalence class of φ with respect to this equivalence relation. We define the *homotopy category* CHCc of chain complexes by taking chain complexes as objects, equivalence classes of morphisms in Cc as morphisms in CHCc, and the formula

$$[\psi] \circ [\varphi] := [\psi\varphi] \tag{3.6}$$

for arbitrary $\psi \in$ Cc$((C', d'), (C'', d''))$ as the definition of composition of morphisms in CHCc. Note that then the equivalence classes of identities in Cc are the identities in CHCc.

Proposition 3.4.2 *The category* CHCc *is well-defined.*

Proof We only verify that the composition (3.6) is well-defined, since the remaining assertions can be established in a straightforward way. Thus, we need to show that if $\varphi, \varphi' : (C, d) \to (C', d')$ and $\psi, \psi' : (C', d') \to (C'', d'')$ are chain homotopic, then the two compositions $\psi\varphi$ and $\psi'\varphi'$ are chain homotopic. For this, let $S : C \to C'$ and $S' : C' \to C''$ be chain homotopies between φ and φ', and between ψ and ψ', respectively. Moreover, consider the map $S'' := \psi' S + S'\varphi$. Then S'' is clearly a degree $+1$ homomorphism, and we have

$$\begin{aligned}\psi'\varphi' - \psi\varphi &= \psi'(\varphi' - \varphi) + (\psi' - \psi)\varphi = \psi'(d'S + Sd) + (d''S' + S'd')\varphi \\ &= \psi'd'S + \psi'Sd + d''S'\varphi + S'd'\varphi \\ &= d''\psi'S + \psi'Sd + d''S'\varphi + S'\varphi d = d''S'' + S''d,\end{aligned}$$

which proves that $\psi\varphi$ and $\psi'\varphi'$ are indeed chain homotopic. □

We have a covariant functor CH : Cc → CHCc which fixes objects and sends a chain map to its chain homotopy equivalence class. Moreover, we say that two chain complexes (C, d) and (C', d') are *chain homotopic* if they are isomorphic in CHCc.

Note that $0_{\mathbb{Z}}$, the $\mathbb{Z}$-graded zero module, together with the zero homomorphism as the boundary map, is a chain complex. We call it the *zero chain complex*. We say that a chain complex (C, d) is *homotopically essential* if it is not chain homotopic to the zero chain complex. Otherwise we say that the chain complex (C, d) is *homotopically trivial* or *homotopically inessential*. Finally, we call a chain complex (C, d) *boundaryless* if $d = 0$.

Proposition 3.4.3 *Assume that (C, d) and (C', d') are two boundaryless chain complexes. Then the chain complexes (C, d) and (C', d') are chain homotopic if and only if C and C' are isomorphic as $\mathbb{Z}$-graded modules.*

Proof Suppose that the chain complexes (C, d) and (C', d') are boundaryless. First assume that C and C' are isomorphic as $\mathbb{Z}$-graded modules. Let $\varphi : C \to C'$ and $\varphi' : C' \to C$ be mutually inverse isomorphisms of $\mathbb{Z}$-graded modules. Since the boundary homomorphisms d and d' are both zero, we have $\varphi d = 0 = d'\varphi$, as well as $\varphi' d' = 0 = d\varphi'$. This in turn implies that both φ and φ' are in fact chain maps, and from $\varphi'\varphi = \mathrm{id}_C$ and $\varphi\varphi' = \mathrm{id}_{C'}$, we obtain $[\varphi'][\varphi] = [\mathrm{id}_C]$ and $[\varphi][\varphi'] = [\mathrm{id}_{C'}]$. This shows that $[\varphi]$ and $[\varphi']$ are mutually inverse isomorphisms in CHCC.

For the reverse implication, we now assume that (C, d) and (C', d') are chain homotopic. Choose chain maps $\varphi : (C, d) \to (C', d')$ and $\varphi' : (C', d') \to (C, d)$ such that $[\varphi] : (C, d) \to (C', d')$ and $[\varphi'] : (C', d') \to (C, d)$ are mutually inverse isomorphisms in CHCC. Let $S : C \to C$ and $S' : C' \to C'$ be the chain homotopies between $\varphi'\varphi$ and id_C, and between $\varphi\varphi'$ and $\mathrm{id}_{C'}$, respectively. Then one has $\mathrm{id}_C - \varphi'\varphi = Sd + dS = 0$ and $\mathrm{id}_{C'} - \varphi\varphi' = S'd' + d'S' = 0$, which proves that φ and φ' are mutually inverse isomorphisms in CC. □

Corollary 3.4.4 *The only chain complex which is both boundaryless and homotopically trivial is the zero chain complex.* □

We now turn our attention to the homology of chain complexes. The *homology module* of a chain complex (C, d) is the $\mathbb{Z}$-graded module $H(C) := (H_n(C, d))_{n\in\mathbb{Z}}$, where we define $H_n(C, d) := \ker d_n / \operatorname{im} d_{n+1}$. In the sequel, we will consider the homology module as a boundaryless chain complex, that is, as a chain complex with zero boundary homomorphism.

By a *homology decomposition* of a chain complex (C, d), we mean a direct sum decomposition $C = V \oplus H \oplus B$ such that V, H, and B are all $\mathbb{Z}$-graded submodules of C, which satisfy both $d_{|H} = 0$ and $d(V) \subset B$, and such that $d_{|V} : V \to B$ is a module isomorphism. Notice that then we also have $d_{|B} = 0$, since $B = d(V)$ implies $d(B) = d^2(V) = \{0\}$. Finally, a *homology complex* of a chain complex (C, d) is defined as a boundaryless chain complex which is chain homotopic to (C, d). Then one has the following proposition.

Proposition 3.4.5 *Assume that R is a field and that the pair (C, d) is a chain complex over R. Then the following statements hold:*

(i) There exists a homology decomposition of (C, d).
(ii) If $C = V \oplus H \oplus B$ is a homology decomposition of the chain complex (C, d), then $(H, 0)$ is chain homotopic to (C, d). Therefore, it is a homology complex of (C, d). Moreover, the modules H and $H(C)$ are isomorphic as $\mathbb{Z}$-graded modules, where $H(C)$ denotes the homology module of C.

Proof To prove statement (i), consider both $Z := \ker d$ and $B := \operatorname{im} d$. Since R is a field, there are $\mathbb{Z}$-graded submodules $V \subset C$ and $H \subset Z$ such that both $C = V \oplus Z$ and $Z = H \oplus B$ are satisfied, and $d_{H\oplus B} = 0$. Clearly, one has $d(V) \subset \operatorname{im} d = B$. We will prove that $d_{|V} : V \to B$ is an isomorphism. Indeed, if $dx = 0$ for some $x \in V$, then we have $x \in Z \cap V = \{0\}$, and hence $d_{|V}$ is a monomorphism. To see that it is also surjective, consider an arbitrary $y \in B$. Since $B = \operatorname{im} d$, we have $y = dx$ for some $x \in C$. In addition, we can write $x = v + z$ for a $v \in V$ and some $z \in Z$. It follows that $dv = dv + dz = dx = y$, which establishes $d_{|V}$ as an epimorphism, and proves statement (i).

To prove (ii), assume that $C = V \oplus H \oplus B$ is a homology decomposition of (C, d). Let $\iota : H \to C$ denote inclusion, and let $\pi : C \to H$ denote the projection map defined via $\pi(v + h + b) = h$ for $v + h + b \in V \oplus H \oplus B = C$.

One can easily see that $\iota : (H, 0) \to (C, d)$ and $\pi : (C, d) \to (H, 0)$ are chain maps and that $\pi\iota = \mathrm{id}_H$. We will show that $\iota\pi$ is chain homotopic to id_C. For this, define the degree $+1$ homomorphism $S : C \to C$ by $Sx = d_{|V}^{-1}b$, where $x = v + h + b \in C = V \oplus H \oplus B$. Then we have

$$\begin{aligned}(dS + Sd)x &= (dS + Sd)(v + h + b) = dd_{|V}^{-1}b + d_{|V}^{-1}dv = b + v \\ &= (b + v + h) - h = x - \iota\pi x = (\mathrm{id}_C - \iota\pi)x,\end{aligned}$$

which proves that S is a chain homotopy joining $\iota\pi$ and id_C. Finally, directly from the homology decomposition definition, one obtains $\ker d = H \oplus B$ and $\operatorname{im} d = B$. Therefore, we have $H(C) = \ker d / \operatorname{im} d \cong H \oplus B/B \cong H$, where both of the isomorphisms are $\mathbb{Z}$-graded. □

In the case of field coefficients, the concepts of homology module and homology complex are essentially the same as the following theorem shows.

Theorem 3.4.6 *Assume that R is a field. Then the following hold:*

(i) Every chain complex admits a homology complex.
(ii) Two chain complexes are chain homotopic if and only if the associated homology complexes are isomorphic as $\mathbb{Z}$-graded modules.
(iii) The homology module of a chain complex is its homology complex.
(iv) Two chain complexes are chain homotopic if and only if the associated homology modules are isomorphic as $\mathbb{Z}$-graded modules.

Proof Property (i) follows immediately from Proposition 3.4.5. In order to prove statement (ii), observe that two chain complexes are chain homotopic if and only if the associated homology complexes are chain homotopic. Therefore, property (ii) follows immediately from Proposition 3.4.3. To prove (iii), consider a chain complex (C, d). Then in view of (i), we may consider a homology complex $(H, 0)$ of (C, d). Since (C, d) and $(H, 0)$ are chain homotopic, by a standard theorem of homology theory [39, §13], the homology modules of (C, d) and $(H, 0)$ are isomorphic as $\mathbb{Z}$-graded modules. Hence, it follows from Proposition 3.4.5(ii) that the homology module of (C, d) is its homology complex. Finally, property (iv) follows readily from (ii) and (iii). □

As an immediate consequence of Theorem 3.4.5 and the definition of homology complex, we obtain the following corollary, which, for general coefficients, may also be derived from the Acyclic Carrier Theorem (see [39, Theorem 13.4]).

Corollary 3.4.7 *The homology module of a chain complex is zero if and only if the chain complex is chain homotopic to zero (homotopically inessential).* □

3.5 Lefschetz Complexes

We end our chapter on preliminaries by recalling some fundamental definitions and facts about *Lefschetz complexes*. Their original definition goes back to Lefschetz, see [29, Chapter III, Section 1, Definition 1.1]. They appear in the literature under various names, in particular as *S-complexes* in [36] and just as *complexes* in [21].

Definition 3.5.1 (Lefschetz Complex) We say that (X, κ) is a *Lefschetz complex* over a ring R if $X = (X_q)_{q \in \mathbb{N}_0}$ is a finite set with $\mathbb{N}_0$-gradation, and $\kappa : X \times X \to R$ is a map such that

$$\kappa(x, y) \neq 0 \quad \Rightarrow \quad x \in X_q, \;\; y \in X_{q-1} \;\text{ for some } q \in \mathbb{N}, \tag{3.7}$$

and for any $x, z \in X$, we have

$$\sum_{y \in X} \kappa(x, y)\kappa(y, z) = 0. \tag{3.8}$$

We refer to the elements of X as *cells*, to $\kappa(x, y)$ as the *incidence coefficient* of the cells x and y, and to κ as the *incidence coefficient map*. We define the *dimension* of a cell $x \in X_q$ as q and denote it by $\dim x$. Whenever the incidence coefficient map is clear from context, we often just refer to X as the Lefschetz complex. We say that (X, κ) is *regular* if for any $x, y \in X$ the incidence coefficient $\kappa(x, y)$ is either zero or it is invertible in R.

Let (X, κ) be a Lefschetz complex. We denote by $C_k(X) := R\langle X_k \rangle$ the free R-module spanned by the set X_k of cells of dimension k for $k \in \mathbb{N}_0$, and let $C_k(X)$ denote the zero module for $k < 0$. Then it is clear that the sum $C(X) := \bigoplus_{k \in \mathbb{Z}} C_k(X)$ is a free $\mathbb{Z}$-graded R-module generated by X. Finally, define the module homomorphism $\partial^\kappa : C(X) \to C(X)$ on generators by

$$\partial^\kappa(x) := \sum_{y \in X} \kappa(x, y) y. \tag{3.9}$$

Then we have the following fundamental result and definition.

Proposition and Definition 3.5.2 *The pair* $(C(X), \partial^\kappa)$ *is a chain complex. We call it the* chain complex of (X, κ) *and refer to the homology of this chain complex as the* Lefschetz homology *of* (X, κ).

Proof Condition (3.7) guarantees that ∂^κ is a degree -1 module homomorphism, and condition (3.8) implies that $(\partial^\kappa)^2 = 0$. □

Note that every finitely generated free chain complex is the chain complex of a Lefschetz complex obtained by selecting a basis. More precisely, assume that (C, ∂) is a finitely generated free chain complex over a ring R and let $U \subset C$ denote a fixed basis of C. Suppose further that $C_k = 0$ for all $k < 0$. Then for every $v \in U$, there are uniquely determined coefficients $a_{vu} \in R$ such that

$$\partial v = \sum_{u \in U} a_{vu} u.$$

Let $\kappa_\partial : U \times U \to R$ be defined by $\kappa_\partial(v, u) = a_{vu}$. Then the following proposition is straightforward.

Proposition 3.5.3 *The pair (U, κ_∂) is a Lefschetz complex.* □

The family of cells of a simplicial complex [23, Definition 11.8] and the family of elementary cubes of a cubical set [23, Definition 2.9] provide simple but important examples of Lefschetz complexes. In these two cases, the respective formulas for the incident coefficients are explicit and elementary, see, for example, [36]. Also a general regular cellular complex, or a regular finite CW complex as considered in [32, Section IX.3], is an example of a Lefschetz complex. In this case, the incident coefficients may be obtained from a system of equations as described in [32, Section IX.5], and the Lefschetz homology may be computed efficiently as outlined in [14]. Note that a Lefschetz complex over a field is always regular.

Given two cells $x, y \in X$, we say that y is a *facet* of x, and we write $y \prec_\kappa x$, if the inequality $\kappa(x, y) \neq 0$ holds. One can easily see that the transitive closure of the relation $\prec_\kappa$ is a partial order. We denote this partial order by $\leq_\kappa$, call it the *face relation*, and denote the associated strict order by $<_\kappa$. As an immediate consequence of (3.7), we have the following proposition.

Proposition 3.5.4 *The map* $\dim : (X, \leq_\kappa) \to (\mathbb{Z}, \leq)$ *that assigns to a cell $x \in X$ its dimension* $\dim x \in \mathbb{Z}$ *is order preserving. Moreover, if the inequality $x \prec_\kappa y$ holds, then one has* $\dim y = \dim x + 1$. □

We say that y is a *face* of x if $y \leq_\kappa x$. The T_0 topology defined via the Alexandrov theorem [1] by the partial order $\leq_\kappa$ is called the *Lefschetz topology* of the Lefschetz complex (X, κ). The Lefschetz topology makes the Lefschetz complex a finite topological space. Notice that in this topology the closure of any set $A \subset X$ consists of all faces of all cells in A.

We say that the subset $A \subset X$ is a *κ-subcomplex* or *Lefschetz subcomplex* of X, if the pair $(A, \kappa_{|A \times A})$ with the $\mathbb{Z}$-gradation induced from X is a Lefschetz complex in its own right. The following proposition provides sufficient conditions for a subset of a Lefschetz complex to be a Lefschetz subcomplex. For more details, see [35, Proposition 5.3] and [36, Theorem 3.1].

Proposition 3.5.5 *If $A \subset X$ is locally closed in the Lefschetz topology, then A is a Lefschetz subcomplex of (X, κ). In particular, every open and every closed subset of a Lefschetz complex X is again a Lefschetz subcomplex.* □

Example 3.5.6 (Nonuniqueness via Subdivision, ▷ 4.3.3) The easiest way to visualize a Lefschetz complex is by presenting it as a κ-subcomplex of a simplicial complex. A sample Lefschetz complex is presented in Fig. 3.1 as a locally closed

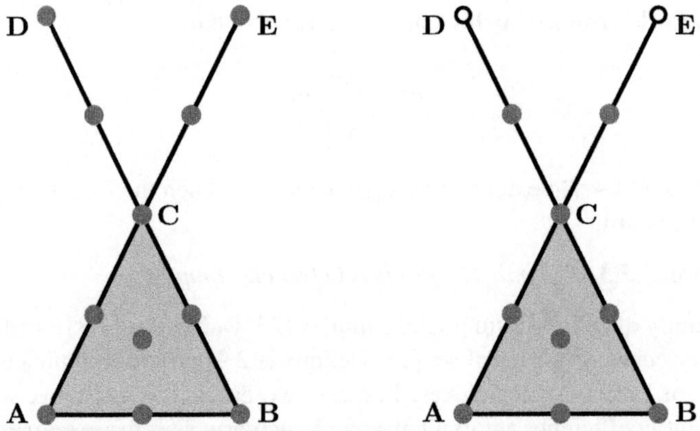

Fig. 3.1 The simplicial complex shown on the left and consisting of the triangle **ABC**, the edges **AB**, **AC**, **BC**, **CD**, **CE**, and the vertices **A**, **B**, **C**, **D**, and **E** is an example of a Lefschetz complex. Simplices are marked with a small circle in the center of mass of each simplex. In the right panel, the simplices marked with a blue circle constitute a locally closed (convex) subset of the set of all simplices. Therefore, it is another example of a Lefschetz complex X which consists of all simplices of the simplicial complex except the vertices **D** and **E**, shown as circles with white interiors

collection of simplices of a simplicial complex. Note, however, that not all Lefschetz complexes can be written in this form. ◊

A Lefschetz subcomplex $A \subset X$ of a Lefschetz complex X has its own Lefschetz topology as well as the associated chain complex $C(A)$. As a topological space A is always a subspace of X, but $C(A)$ typically is not a chain subcomplex of X unless A is closed in X, as the following result shows.

Proposition 3.5.7 *(see [35, Theorem 5.4]) If A is closed in X in the Lefschetz topology, then*

$$\partial^{\kappa|A\times A} = \partial^{\kappa}_{|C(A)} \tag{3.10}$$

and

$$\partial^{\kappa}(C(A)) \subset C(A). \tag{3.11}$$

In particular, the chain complex $(C(A), \partial^{\kappa|A\times A})$ is a chain subcomplex of the chain complex $(C(X), \partial^{\kappa})$.

Proof To see (3.10), it suffices to verify the equality on basis elements. Thus, take an $x \in A$. Then

$$\partial^{\kappa|A\times A} x = \sum_{y\in A} \kappa(x, y)y = \sum_{y\in X} \kappa(x, y)y = \partial^{\kappa} x,$$

because $\kappa(x, y) \neq 0$ implies $y \in \operatorname{cl} x \subset \operatorname{cl} A = A$. Property (3.11) and the remaining assertion are obvious. □

We close with a few comments about the homology of Lefschetz complexes. Given a closed subset $A \subset X$ in the Lefschetz topology, we define the *relative Lefschetz homology* $H(X, A)$ as the homology of the quotient chain complex $(C(X, A), \tilde{\partial})$, where we define $C(X, A) := C(X)/C(A)$, and $\tilde{\partial}$ stands for the induced boundary map. We then have the following proposition, which follows immediately from [35, Theorem 5.4] (see also [36, Theorem 3.5]), and which uses relative Lefschetz homology to characterize the Lefschetz homology of a locally closed subset $A \subset X$.

Proposition 3.5.8 *If $A \subset X$ is locally closed in the Lefschetz topology of the Lefschetz complex X, then the Lefschetz homology $H(A)$ is isomorphic to the relative homology $H(\operatorname{cl} A, \operatorname{mo} A)$ of the pair $(\operatorname{cl} A, \operatorname{mo} A)$ of closed subsets of X.*

□

The following result computes the homology of some simple Lefschetz complexes.

Proposition 3.5.9 *(see [35, Proposition 5.2]) For every $x \in X$, the singleton $\{x\}$ is a Lefschetz subcomplex of X, and its Lefschetz homology is given by*

$$H_q(\{x\}) \cong \begin{cases} R & \text{if } q = \dim x, \\ 0 & \text{otherwise.} \end{cases}$$

For every $x, y \in X$ such that x is a facet of y, the doubleton $\{x, y\}$ is a Lefschetz subcomplex of X, and for all $q \in \mathbb{Z}$, one has $H_q(\{x, y\}) \cong 0$. □

Chapter 4
Poset Filtered Chain Complexes

In the previous chapter, we recalled basic definitions and properties of chain complexes and their homology. For the definition of the connection matrix, these considerations have to be extended to the case of chain complexes which are poset filtered. In addition, we need to study maps between these poset filtered chain complexes, even in the situation where the filtration posets of the involved chain complexes differ. All of these concepts will be introduced in the current chapter.

4.1 Graded and Filtered Module Homomorphisms

We begin our presentation on the level of general graded modules. Consider a K-graded module M and a K'-graded module M'. Furthermore, let $\alpha : K' \nrightarrow K$ be a partial map. Then we say that a homomorphism $h : M \to M'$ is *α-graded* if for every pair of indices $j \in K$ and $i \in K'$ we have

$$h_{ij} \neq 0 \quad \Rightarrow \quad i \in \alpha^{-1}(j),$$

which is equivalent to the implication

$$h_{ij} \neq 0 \quad \Rightarrow \quad i \in \operatorname{dom} \alpha \quad \text{and} \quad \alpha(i) = j. \tag{4.1}$$

If in addition the sets K and K' are posets and α is order preserving, then we say that a homomorphism $h : M \to M'$ is *α-filtered* if for every $j \in K$ and $i \in K'$ one has

$$h_{ij} \neq 0 \quad \Rightarrow \quad i \in \alpha^{-1}(j^{\leq})^{\leq}, \tag{4.2}$$

M. Mrozek, T. Wanner, *Connection Matrices in Combinatorial Topological Dynamics*, SpringerBriefs in Mathematics,
https://doi.org/10.1007/978-3-031-87600-4_4

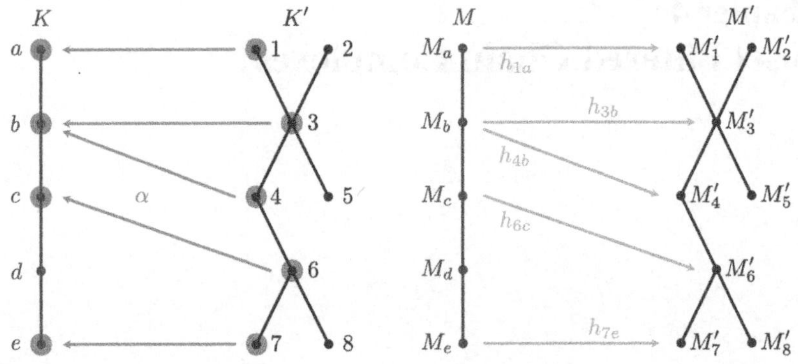

Fig. 4.1 Sample α-graded morphism. The left image shows two posets K and K', together with an order preserving partial map $\alpha : K' \nrightarrow K$. The panel on the right indicates the potential nonzero module homomorphisms h_{ij} if h is α-graded

which is equivalent to

$$h_{ij} \neq 0 \quad \Rightarrow \quad \exists_{i' \in \operatorname{dom} \alpha} \; i \leq i' \quad \text{and} \quad \alpha(i') \leq j.$$

These two definitions are illustrated in Figs. 4.1 and 4.2. The left image in Fig. 4.1 shows two partially ordered sets together with a partial map α between them. In the right image of the same figure, arrows indicate which of the homomorphisms h_{ij} can be nontrivial if h is α-graded. In contrast, Fig. 4.2 depicts all possible nontrivial homomorphisms h_{ij} if h is α-filtered. The more involved implication (4.2) is illustrated in the top left image, which indicates the sets $3^{\leq}$ and $b^{\leq}$ in light blue. The remaining three panels in the figure depict all possible nontrivial homomorphisms h_{ij}. It is clear from this example that α-filtered homomorphisms are far less restricted than α-graded ones. In fact, an α-filtered homomorphism can have a nontrivial h_{ij} even if $i \notin \operatorname{dom} \alpha$ or if $j \notin \operatorname{im} \alpha$.

Conditions (4.1) and (4.2) for α-graded and α-filtered homomorphisms, respectively, may be simplified if $\operatorname{dom} \alpha = K'$, i.e., if α is actually a function. This is the subject of the following straightforward proposition.

Proposition 4.1.1 *Assume that* $\operatorname{dom} \alpha = K'$. *Then* $h : M \to M'$ *is* α*-graded if and only if for every* $i \in K'$ *and* $j \in K$

$$h_{ij} \neq 0 \quad \Rightarrow \quad \alpha(i) = j, \tag{4.3}$$

and h *is* α*-filtered if and only if*

$$h_{ij} \neq 0 \quad \Rightarrow \quad \alpha(i) \leq j \tag{4.4}$$

for every $i \in K'$ *and* $j \in K$. □

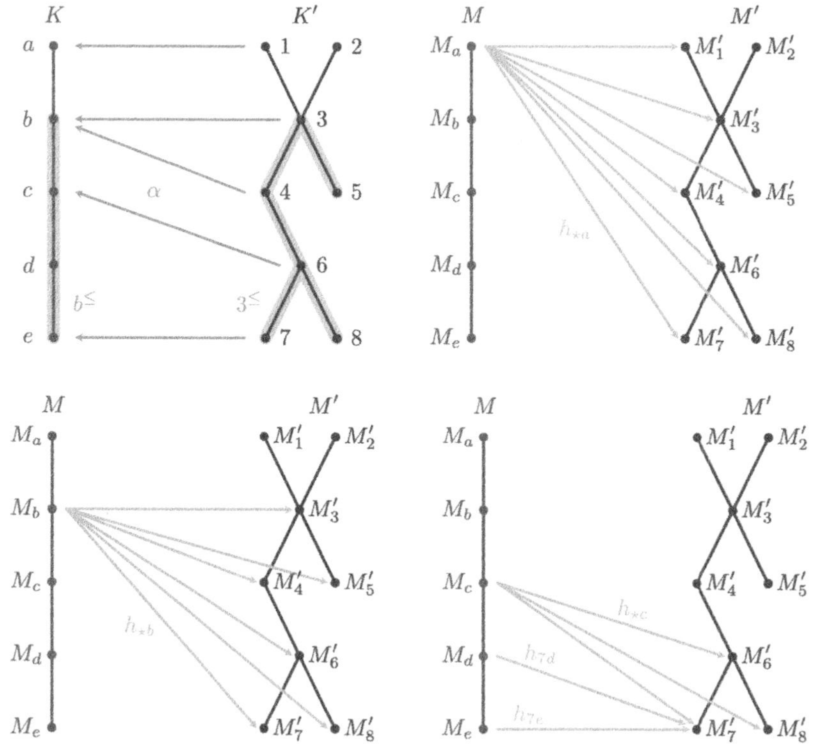

Fig. 4.2 Sample α-filtered morphism. The top left image shows two posets K and K', together with an order preserving partial map $\alpha : K' \nrightarrow K$. The remaining three panels indicate all potentially nonzero module homomorphisms h_{ij} if h is α-filtered (top right for $j = a$, bottom left for $j = b$ and bottom right for $j = c, d, e$)

Obviously, if α is order preserving and h is an α-graded homomorphism, then h is an α-filtered homomorphism. However, the converse is not true in general. This can readily be seen from Figs. 4.1 and 4.2.

In the case when $K = K'$, i.e., when the two posets are the same, we say that a homomorphism $h : M \to M'$ is *K-filtered* (respectively, *K-graded*) if h is id_K-filtered (respectively, id_K-graded). We simplify the terminology to *filtered* (respectively, *graded*) if K is clear from context. As an immediate consequence of Proposition 4.1.1, we get the following corollary.

Corollary 4.1.2 *An endomorphism $h : M \to M$ on a module M is graded if and only if for every $i, j \in K$ we have*

$$h_{ij} \neq 0 \quad \Rightarrow \quad i = j,$$

and h is filtered if and only if one has

$$h_{ij} \neq 0 \quad \Rightarrow \quad i \leq j$$

for every $i, j \in K$. □

Note that in the special case when $K = K' = \mathbb{Z}$, an α-graded homomorphism with respect to the *left shift by k map* $\alpha : \mathbb{Z} \ni i \mapsto i - k \in \mathbb{Z}$ on the integers coincides with a $\mathbb{Z}$-graded homomorphism of degree k. We recall that in such a case, when k is clear from the context, we shorten the notation $h_{j+k,j}$ to h_j.

Proposition 4.1.3 *Assume that* P *and* P' *are posets and that* $\alpha : P' \twoheadrightarrow P$ *is order preserving. Moreover, let* M *be a* P*-graded module and let* M' *denote a* P'*-graded module. Then a module homomorphism* $h : M \to M'$ *is an* α*-filtered homomorphism if and only if*

$$h(M_L) \subset M'_{\alpha^{-1}(L)^{\leq}} \tag{4.5}$$

for every $L \in \mathrm{Down}(P)$.

Proof Assume that $h : M \to M'$ is a module homomorphism such that (4.5) is satisfied for every $L \in \mathrm{Down}(P)$. In addition, suppose that $h_{pq} \neq 0$ for $q \in P$ and $p \in P'$. Let $x \in M_q$ be such that $h_{pq}(x) \neq 0$. If we consider $L := q^{\leq} \in \mathrm{Down}(P)$, then (4.5) implies $h(M_L) \subset M'_{\alpha^{-1}(L)^{\leq}}$. Since we have the inclusions $x \in M_q \subset M_L$, we obtain $h(x) \in M'_{\alpha^{-1}(L)^{\leq}}$. Thus, the identity $h_{p'q}(x) = \pi_{p'}(h(x)) = 0$ holds as long as $p' \notin \alpha^{-1}(L)^{\leq}$, which in turn implies the inclusion $p \in \alpha^{-1}(L)^{\leq}$, in view of the fact that $h_{pq}(x) = \pi_p(h(x)) \neq 0$. Therefore, we have $p \leq \bar{p}$ for an element $\bar{p}$ with $\alpha(\bar{p}) \in L = q^{\leq}$. Finally, one has $p \in \alpha^{-1}(q^{\leq})^{\leq}$, which implies (4.2) and proves that h is α-filtered.

To verify the opposite implication, assume that $h : M \to M'$ is a module homomorphism such that (4.2) holds. Let $L \in \mathrm{Down}(P)$ and consider $x \in M_L$. Without loss of generality, we may assume that $x \in M_q$ for a $q \in L$. Then (4.2) yields

$$h(x) = \sum_{p \in P} h_{pq}(x) = \sum_{p \in \alpha^{-1}(q^{\leq})^{\leq}} h_{pq}(x),$$

which means that $h(x) \in M'_{\alpha^{-1}(L)^{\leq}}$, since we have $\alpha^{-1}(q^{\leq})^{\leq} \subset \alpha^{-1}(L)^{\leq}$. This establishes the inclusion in (4.5). □

In the special case when $P = P'$ and $\alpha = \mathrm{id}_P$, we have the following corollary of Proposition 4.1.3.

Corollary 4.1.4 *Assume that P is a poset and that M is a P-graded module. Then the homomorphism $h : M \to M$ is a filtered homomorphism if and only if*

$$h(M_L) \subset M_L \tag{4.6}$$

for every $L \in \mathrm{Down}(P)$.

Proof By applying our assumptions $P = P'$ and $\alpha = \mathrm{id}_P$ to the setting of Proposition 4.1.3, one obtains that $\alpha^{-1}(L)^{\leq} = L^{\leq} = L$. Therefore, the condition in (4.5) reduces to the one in (4.6). □

Definition 4.1.5 We say that two P-gradations $(M_p)_{p\in P}$ and $(M'_p)_{p\in P}$ of the same module M are *filtered equivalent*, if for every $L \in \mathrm{Down}(P)$ we have $M_L = M'_L$.

As an immediate consequence of Corollary 4.1.4, one obtains the following.

Proposition 4.1.6 *Let P be a poset. Assume that M and M' are P-graded modules and that $h : M \to M'$ is a filtered homomorphism. Then h is a filtered homomorphism with respect to any filtered equivalent gradation of the modules M and M'.* □

4.2 The Category of Graded and Filtered Moduli

We define the two categories GMod of graded moduli and FMod of filtered moduli as follows. The objects of GMod (respectively, FMod) are pairs (P, M) where P is an object of DSet (respectively, DPSet) and M is a P-graded module (see Sect. 3.3). The morphisms in GMod (respectively, FMod) are pairs $(\alpha, h) : (P, M) \to (P', M')$ such that $\alpha : P' \nrightarrow P$ is a morphism in DSet (respectively, DPSet) and $h : M \to M'$ is an α-graded (respectively, α-filtered) module homomorphism. In the following, we will briefly refer to morphisms in GMod (respectively, FMod) as graded (respectively, filtered) morphisms $(\alpha, h) : (P, M) \to (P', M')$. In the special case when one has the identities $(P, M) = (P', M')$ and $\alpha = \mathrm{id}_P$, we simplify the terminology by referring to h as a filtered (respectively, graded) morphism, which of course means that the homomorphism h is id_P-filtered (respectively, id_P-graded).

One can easily verify that $\mathrm{id}_{(P,M)} := (\mathrm{id}_P, \mathrm{id}_M)$ is the identity morphism on (P, M) in both the GMod and FMod categories. If $(\alpha, h) : (P, M) \to (P', M')$ and $(\alpha', h') : (P', M') \to (P'', M'')$ are two graded or filtered morphisms, then we define their composition

$$(\alpha'', h'') := (\alpha', h') \circ (\alpha, h) : (P, M) \to (P'', M'')$$

by setting $\alpha'' := \alpha\alpha'$ and $h'' := h'h$. This leads to the following fundamental result.

Proposition 4.2.1 GMod *is a well-defined category.*

Proof The only nontrivial part of the proof is the verification that the composition of two graded morphisms is again a graded morphism. For this, let (α'', h'') be given as above. Obviously, the composition $h'h$ is a module homomorphism. We will show that $h''_{pr} \neq 0$ for $p \in P''$ and $r \in P$ implies the equality $(\alpha\alpha')(p) = r$. Hence, we assume that $h''_{pr} \neq 0$ for some $p \in P''$ and $r \in P$. It follows from Proposition 3.3.3 that there exists a $q \in P'$ such that both $h'_{pq} \neq 0$ and $h_{qr} \neq 0$ are satisfied. This further shows that one has to have $p \in \operatorname{dom} \alpha'$ and $\alpha'(p) = q$, as well as $q \in \operatorname{dom} \alpha$ and $\alpha(q) = r$. It follows that $p \in \operatorname{dom}(\alpha\alpha')$ and $(\alpha\alpha')(p) = r$. This proves that the pair (α'', h'') is indeed a graded morphism. □

Suppose that $f : (P, M) \to (P, M)$ and $f' : (P', M') \to (P', M')$ are two poset graded morphisms. They are called *graded-conjugate*, if there exists a poset graded isomorphism $(\alpha, h) : (P, M) \to (P', M')$ such that

$$(\alpha, h) \circ f = f' \circ (\alpha, h).$$

If f and f' are graded-conjugate, then we say that their (P, P)- and (P', P')-matrices are *graded similar*. It can be easily seen that graded similarity of matrices is an equivalence relation.

Proposition 4.2.2 FMOD *is a well-defined category.*

Proof We only need to verify that the composition of filtered morphisms is again filtered. For this, consider a composition $(\alpha'', h'') := (\alpha', h') \circ (\alpha, h)$ of two filtered morphisms as above. Obviously, the map $\alpha\alpha'$ is order preserving, and $h'h$ is a module homomorphism. Thus, we only need to verify that $h''_{pr} \neq 0$ for $p \in P''$ and $r \in P$ implies the inclusion $p \in (\alpha\alpha')^{-1}(r^{\leq})^{\leq}$. Hence, assume that $h''_{pr} \neq 0$ for $p \in P''$ and $r \in P$. It follows from Proposition 3.3.3 that there exists a $q \in P'$ such that one has both $h'_{pq} \neq 0$ and $h_{qr} \neq 0$. Then there exist $\bar{p} \geq p$ and $\bar{q} \geq q$ such that $\alpha'(\bar{p}) \leq q$ and $\alpha(\bar{q}) \leq r$. Since α is order preserving, it follows that $(\alpha\alpha')(\bar{p}) \leq \alpha(\bar{q}) \leq r$. Hence, $\bar{p} \in (\alpha\alpha')^{-1}(r^{\leq})$ and $p \in (\alpha\alpha')^{-1}(r^{\leq})^{\leq}$. This proves that (α', h'') is indeed a filtered morphism. □

Proposition 4.2.3 *Suppose we are given morphisms* $(\alpha, h) : (P, M) \to (P', M')$ *and* $(\alpha', h') : (P', M') \to (P'', M'')$ *in* FMOD *which satisfy both* $\operatorname{dom} \alpha = P'$ *and* $\operatorname{dom} \alpha' = P''$. *Assume in addition that* α *is injective. If* $p \in P$ *and* $p'' \in P''$ *are such that* $\alpha\alpha'(p'') = p$, *then we have*

$$(h'h)_{p''p} = h'_{p''\alpha'(p'')} h_{\alpha'(p'')p} \quad \textit{for } p \in P. \tag{4.7}$$

Proof We get from Proposition 3.3.3 that

$$(h'h)_{p''p} = \sum_{p' \in P'} h'_{p''p'} h_{p'p}. \tag{4.8}$$

It follows from (4.4) that the index p' of a nonzero term on the right-hand side of (4.8) has to satisfy both $\alpha'(p'') \leq p'$ and $\alpha(p') \leq p$. Hence,

$$p = \alpha\alpha'(p'') \leq \alpha(p') \leq p.$$

It follows that $\alpha(p') = p$, and, since α is injective, we get $p' = \alpha'(p'')$. This proves the equality in (4.7). □

As an immediate consequence of Proposition 4.2.3, one obtains the following corollary.

Corollary 4.2.4 *Suppose we are given two morphisms $(\alpha, h) : (P, M) \to (P', M')$ and $(\alpha', h') : (P', M') \to (P, M)$ in* FMOD *such that $\alpha : P' \to P$ and $\alpha' : P \to P'$ are mutually inverse bijections. Then*

$$(h'h)_{pp} = h'_{p\alpha'(p)} h_{\alpha'(p)p} \quad \textit{for } p \in P.$$

□

Lemma 4.2.5 *A filtered morphism $(\alpha, h) : (P, M) \to (P', M')$ is an isomorphism in* FMOD *if and only if the map $\alpha : P' \to P$ is an order isomorphism between posets, and the map $h_{p'\alpha(p')} : M_{\alpha(p')} \to M'_{p'}$ is a module isomorphism for every $p' \in P'$.*

Proof We begin by assuming that (α, h) is an isomorphism in the category FMOD and that $(\alpha', h') : (P', M') \to (P, M)$ is its inverse. Then $\alpha\alpha' = \mathrm{id}_P$ and $\alpha'\alpha = \mathrm{id}_{P'}$. Since α and α' are order preserving, we see that $\alpha : P' \to P$ is an order isomorphism. It follows from Corollary 4.2.4 that

$$\mathrm{id}_{M_p} = (\mathrm{id}_M)_{pp} = (h'h)_{pp} = h'_{p\alpha'(p)} h_{\alpha'(p)p} \text{ for } p \in P \tag{4.9}$$

and

$$\mathrm{id}_{M_{p'}} = (\mathrm{id}_M)_{p'p'} = (hh')_{p'p'} = h_{p'\alpha(p')} h'_{\alpha(p')p'} \text{ for } p' \in P'. \tag{4.10}$$

However, the two maps α and α' are mutually inverse bijections. If we therefore use the substitution $p' := \alpha'(p)$ in (4.9), we get $p = \alpha(p')$, as well as

$$\mathrm{id}_{M_{\alpha(p')}} = h'_{\alpha(p')p'} h_{p'\alpha(p')} \text{ for } p' \in P'.$$

It finally follows that $h'_{\alpha(p')p'}$ is the inverse of $h_{p'\alpha(p')}$. Thus, the map $h_{p'\alpha(p')}$ is a module isomorphism for every $p' \in P'$.

To verify the opposite implication, assume that $(\alpha, h) : (P, M) \to (P', M')$ is a filtered homomorphism such that $\alpha : P' \to P$ is an order isomorphism, and the map $h_{p'\alpha(p')} : M_{\alpha(p')} \to M'_{p'}$ is a module isomorphism for all $p' \in P'$. Since P and P' as objects of DPSET are finite sets, we may proceed by induction on their cardinality $n := \operatorname{card} P = \operatorname{card} P'$. If $n = 1$, then $P = \{p\}$, $P' = \{p'\}$, and $h_{p'\alpha(p')} = h_{p'p} = h$ is a module isomorphism. Clearly, its inverse is a filtered homomorphism. Hence, h is an isomorphism in FMOD. Thus, we now assume $n > 1$. Let $\bar{p}'$ be a maximal element in P'. Set $\bar{p} := \alpha(\bar{p}')$. Since α' is an order isomorphism, we see that $\bar{p}$ is a maximal element in P. Let $\bar{P}' := P' \setminus \{\bar{p}'\}$, $\bar{P} := P \setminus \{\bar{p}\}$, as well as $\bar{M}' := M_{\bar{P}'}$, $\bar{M} := M_{\bar{P}}$. Let $Q := \{\bar{P}, \bar{p}\}$ and $Q' := \{\bar{P}', \bar{p}'\}$ be linearly ordered, respectively, by $\bar{P} < \bar{p}$ and $\bar{P}' < \bar{p}'$. Then (Q, M) and (Q', M') are graded modules. Define $\bar{\alpha} : Q' \to Q$ through the identities $\bar{\alpha}(\bar{P}') := \bar{P}$ and $\bar{\alpha}(\bar{p}') := \bar{p}$. We will prove

$$h_{\bar{p}'p} = 0 \text{ for } p \in \bar{P}. \tag{4.11}$$

Arguing by contradiction, assume that $h_{\bar{p}'p} \neq 0$ for an element $p \in \bar{P}$. Since h is an α-filtered homomorphism, we get from (4.2) that $h_{\bar{p}'p} \neq 0$ implies $\bar{p}' \leq \bar{\bar{p}}'$ for some $\bar{\bar{p}}' \in \operatorname{dom} \alpha$ such that $\alpha(\bar{\bar{p}}') \leq p$. Since $\bar{p}'$ is maximal in P', we obtain both $\bar{p}' = \bar{\bar{p}}'$ and $\bar{p} = \alpha(\bar{p}') \leq p$. Since $\bar{p}$ is maximal in P, this further implies the equality $p = \bar{p}$, a contradiction proving (4.11). Hence, the identity in (3.4) yields the identity $h_{\bar{p}'\bar{P}} = \sum_{p \in \bar{P}} h_{\bar{p}'p} \circ \pi_p = 0$. Therefore, the (Q, Q')-matrix of h is

$$\begin{bmatrix} h_{\bar{P}'\bar{P}} & h_{\bar{P}'\bar{p}} \\ 0 & h_{\bar{p}'\bar{p}} \end{bmatrix}.$$

By induction assumption, the pair $(\alpha_{|\bar{P}'}, h_{\bar{P}'\bar{P}}) : M_{\bar{P}} \to M_{\bar{P}'}$ is an isomorphism in FMOD. Set $\alpha' := \alpha^{-1}$ and let $h' : M' \to M$ be the module homomorphism given by the (Q', Q)-matrix

$$\begin{bmatrix} h^{-1}_{\bar{P}'\bar{P}} & -h^{-1}_{\bar{P}'\bar{P}} h_{\bar{P}'\bar{p}} h^{-1}_{\bar{p}'\bar{p}} \\ 0 & h^{-1}_{\bar{p}'\bar{p}} \end{bmatrix}.$$

One easily verifies that the homomorphism h' is α'-filtered, and a straightforward computation shows that $(\alpha', h') \circ (\alpha, h) = \mathrm{id}_{(P,M)}$ and $(\alpha, h) \circ (\alpha', h') = \mathrm{id}_{(P',M')}$. Hence, the morphism (α, h) is indeed an isomorphism in FMOD. □

Corollary 4.2.6 *Assume that the pair $(\alpha, h) : (P, M) \to (P', M')$ is both a homomorphism in* GMOD *and an isomorphism in* FMOD. *Then (α, h) is automatically an isomorphism in* GMOD.

Proof In view of Lemma 4.2.5, the map $\alpha : P' \to P$ is an order isomorphism, and the map $h_{p'\alpha(p')} : M_{\alpha(p')} \to M'_{p'}$ is a module isomorphism. Let $g : M' \to M$ be the α^{-1}-graded homomorphism with (P', P)-matrix given by

$$g_{pp'} := \begin{cases} h^{-1}_{p'\alpha(p')} & \text{if } p = \alpha(p'), \\ 0 & \text{otherwise.} \end{cases}$$

Then it is straightforward to verify that (α^{-1}, g) is the inverse of (α, h) in GMOD. Hence, the pair (α, h) is an isomorphism in GMOD. □

4.3 Poset Filtered Chain Complexes

Let P be an arbitrary finite set and let (C, d) be a chain complex. We call (C, d) a *P-graded chain complex*, if C is a P-graded module in which each $C_p \subset C$ is a $\mathbb{Z}$-graded submodule of C. Since the chain complex C is P-graded, its boundary homomorphism d has a (P, P)-matrix. A partial order $\leq$ in P is called (C, d)-admissible, or briefly *d-admissible*, if d is a filtered homomorphism with respect to $(P, \leq)$, i.e., if we have

$$d_{pq} \neq 0 \Rightarrow p \leq q \tag{4.12}$$

for all $p, q \in P$. Note that a (C, d)-admissible partial order on P may not always exist. However, we have the following straightforward proposition and definition.

Proposition and Definition 4.3.1 *If a given P-graded chain complex (C, d) admits a (C, d)-admissible partial order on P, then the intersection of all such (C, d)-admissible partial orders on P is again a (C, d)-admissible partial order on P. We call it the* native *partial order of d.* □

Definition 4.3.2 (Poset Filtered Chain Complex) We say that the triple (P, C, d) is a *poset filtered chain complex* if P is a poset, the pair (C, d) is a P-graded chain complex, and the partial order in P is d-admissible. In addition, we will consider the poset P as an object of DPSET with a distinguished subset $P_\star$, which will have to satisfy the additional condition (4.16) described later in this section. For now, the reader can just assume that $P_\star = P$.

Note that for a poset filtered chain complex (P, C, d), the module C does not only have to be P-graded, but also $\mathbb{Z}$-graded, where the n-th summand of the $\mathbb{Z}$-gradation is the direct sum over $p \in P$ of the n-th summands in the $\mathbb{Z}$-gradation of C_p.

Example 4.3.3 (Nonuniqueness via Subdivision, ◁ 3.5.6 ▷ 5.1.2) Motivated by the Lefschetz complex shown in the right panel of Fig. 3.1, we consider the set of words

$$X := \{\mathbf{A}, \mathbf{B}, \mathbf{C}, \mathbf{AB}, \mathbf{AC}, \mathbf{BC}, \mathbf{CD}, \mathbf{CE}, \mathbf{ABC}\}$$

and the free module $C := \mathbb{Z}_2\langle X\rangle$ with basis X and coefficients in the field $\mathbb{Z}_2$. For a word $x \in X$, define its *dimension* as one less than the number of characters in x and denote the set of words of dimension i by X_i. Setting

$$C_i := \begin{cases} \mathbb{Z}_2\langle X_i \rangle & \text{if } X_i \neq \varnothing, \\ 0 & \text{otherwise,} \end{cases}$$

for $i \in \mathbb{Z}$, we obtain a $\mathbb{Z}$-gradation of C given by

$$C = \bigoplus_{i \in \mathbb{Z}} C_i. \tag{4.13}$$

In addition, let $P = \{\mathbf{p}, \mathbf{q}, \mathbf{r}, \mathbf{s}, \mathbf{t}, \mathbf{u}\}$ be a poset with its partial order $\leq$ defined by the Hasse diagram

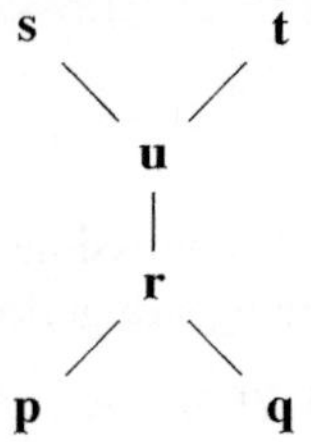

(4.14)

For $p = \mathbf{p}, \mathbf{q}, \mathbf{r}, \mathbf{s}, \mathbf{t}, \mathbf{u} \in P$, we define X_p, respectively, as

$$\{\mathbf{A}\},\ \{\mathbf{B}\},\ \{\mathbf{AB}\},\ \{\mathbf{CD}\},\ \{\mathbf{CE}\},\ \{\mathbf{C}, \mathbf{AC}, \mathbf{BC}, \mathbf{ABC}\}.$$

Then $\{X_p\}_{p \in P}$ is a partition of X which induces a gradation

$$C = \bigoplus_{p \in P} C_p \tag{4.15}$$

with $C_p := \mathbb{Z}_2\langle X_p \rangle$. Consider the homomorphism $d : C \to C$ of degree -1 which is defined on the basis X by the matrix

	A	**B**	**AB**	**C**	**AC**	**BC**	**ABC**	**CD**	**CE**
A			1		1				
B			1			1			
AB							1		
C					1	1		1	1
AC							1		
BC							1		
ABC									
CD									
CE									

.

One can immediately verify that d is P-filtered and satisfies $d^2 = 0$. Therefore, the triple (P, C, d) with the $\mathbb{Z}$-gradation (4.13) and the P-gradation (4.15) is a well-defined poset filtered chain complex. ◊

We would like to point out that if $J \subset P$, then (C_J, d_{JJ}) does not need to be a chain complex in general. However, we have the following proposition.

Proposition 4.3.4 *Assume that J is a convex subset of P. Then we have:*

(i) *(C_J, d_{JJ}) is a chain complex.*
(ii) *(C_J, d_{JJ}) is chain isomorphic to the quotient complex $(C_{J^{\leq}}/C_{J^{<}}, d')$, where d' denotes the homomorphism induced by $d_{J^{\leq}J^{\leq}}$. Notice that the quotient complex is well-defined since $J^{<}$ is a down set due to the convexity of J.*

Proof In order to prove (i), we need to verify that $d_{JJ}^2 = 0$. Assume first that J is a down set. Let $x \in C_J$. Then Corollary 4.1.4 immediately implies that $dx \in C_J$. Hence, one obtains both $d_{JJ}x = (\pi_J \circ d \circ \iota_J)(x) = dx$ and $d_{JJ}^2 x = d_{JJ}dx = d^2x = 0$, which yields $d_{JJ}^2 = 0$. If J is just convex, we consider the down sets $I := J^{<}$ and $K := J^{\leq}$, which clearly satisfy the identity $K = I \cup J$. Since d is a filtered homomorphism, we have $d_{pq} = 0$ for $p \in J$ and $q \in I$. Therefore, the matrix of d_{KK} takes the form

$$\begin{bmatrix} d_{II} & d_{IJ} \\ 0 & d_{JJ} \end{bmatrix}.$$

Since K is a down set, we already have verified that $d_{KK}^2 = 0$. It follows that

$$0 = d_{KK}^2 = \begin{bmatrix} d_{II}^2 & d_{II}d_{IJ} + d_{IJ}d_{JJ} \\ 0 & d_{JJ}^2. \end{bmatrix}$$

Thus, one obtains $d_{JJ}^2 = 0$, which proves statement (i).

To establish (ii), we consider again the down sets $I := J^{<}$ and $K := J^{\leq}$, as well as the homomorphism $\kappa : C_J \ni x \mapsto [x]_I \in C_K/C_I$, where $[x]_I$ denotes the equivalence class of x in the quotient module C_K/C_I. Let $x \in C_J$ and let $\bar{d}_{KK}$ denote the boundary homomorphism induced by d_{KK} on the quotient module C_K/C_I. We then have

$$\bar{d}_{KK}\kappa x = [dx]_I = [d_{IJ}x + d_{JJ}x]_I = [d_{JJ}x]_I = \kappa d_{JJ}x,$$

and this implies that the map κ is a chain map. Assume now that one has $[x]_I = 0$ for an $x \in C_J$. Then $x \in C_I \cap C_J = \{0\}$, that is, $x = 0$. This proves that κ is a monomorphism. Finally, given the class $[y]_I \in C_K/C_I$ generated by $y \in C_K$, we have $y = y_I + y_J$ for a $y_I \in C_I$ and a $y_J \in C_J$. It follows that $[y]_I = [y_J]_I = \kappa y_J$, proving that κ is an epimorphism. Hence, the map κ is an isomorphism. □

It is straightforward to observe that the chain complex (C_J, d_{JJ}) in the above Proposition 4.3.4(i) is in fact J-filtered. Therefore, we have the following proposition.

Proposition 4.3.5 *Consider a filtered chain complex (P, C, d), as well as a convex subset $J \subset P$. Then the triple (J, C_J, d_{JJ}) is also a filtered chain complex. We call it the* filtered chain complex induced by a convex subset. □

Since a singleton $\{p\} \subset P$ is always a convex subset, one also immediately obtains the following corollary.

Corollary 4.3.6 *Let (P, C, d) be an arbitrary poset filtered chain complex. Then the pair (C_p, d_{pp}) is a chain complex for every $p \in P$.* □

Given a poset filtered chain complex (P, C, d), we consider the poset P as an object of DPSET with the distinguished subset $P_\star$ satisfying

$$\{\, p \in P \mid C_p \text{ is homotopically essential} \,\} \subset P_\star. \tag{4.16}$$

Recall that C_p is homotopically essential if it is not chain homotopic to the zero chain complex.

In purely algebraic terms, equality in (4.16) suffices, because, as we discussed in Sect. 2.4, a homotopically inessential C_p contributes nothing to the Conley complex and the connection matrix. In terms of applications, however, it may be difficult to ensure equality in (4.16). In applications to dynamics, the set $P_\star$ comes from a Morse decomposition, and often only inclusion in (4.16) may be guaranteed. In this context, lack of equality in (4.16) can mean that some Morse sets may only be known via isolating neighborhoods and potentially may have zero Conley index.

We would like to point out that in Example 4.3.3 the chain complex (C_p, d_{pp}) is essential for $p = \mathbf{p}, \mathbf{q}, \mathbf{r}, \mathbf{s}, \mathbf{t}$ and inessential for $p = \mathbf{u}$. Therefore, for the poset P in Example 4.3.3, there are only two possible choices of $P_\star$, namely $P_\star = \{\mathbf{p}, \mathbf{q}, \mathbf{r}, \mathbf{s}, \mathbf{t}\}$, and the case $P_\star = P$.

4.4 The Category of Filtered and Graded Chain Complexes

As our next step, we study morphisms between poset filtered chain complexes. For this, consider two poset filtered chain complexes (P, C, d) and (P', C', d'). Then we say that the map $(\alpha, h) : (P, C, d) \to (P', C', d')$ is a *filtered chain morphism*, if $h : (C, d) \to (C', d')$ is a chain map and h is α-filtered, that is, the map $(\alpha, h) : (P, C) \to (P', C')$ is a morphism in FMOD. We say that the map (α, h) is a *graded chain morphism*, if the map $h : (C, d) \to (C', d')$ is a chain map and h is α-graded, that is, the map $(\alpha, h) : (P, C) \to (P', C')$ is a morphism in GMOD.

We define the category PFCC of poset filtered chain complexes by taking poset filtered chain complexes as objects and filtered chain morphisms as morphisms. One easily verifies that this is indeed a category. We also define the subcategory PGCC of poset graded chain complexes by taking the same objects as in PFCc and graded chain morphisms as morphisms.

Definition 4.4.1 Let $(\alpha, \varphi), (\alpha', \varphi') : (P, C, d) \to (P', C', d')$ denote a pair of filtered chain morphisms, and let $(\gamma, \Gamma) : (P, C) \to (P', C')$ be a filtered module morphism. Then the pair (γ, Γ) is called an *elementary filtered chain homotopy* between (α, φ) and (α', φ') if the following three conditions are satisfied:

(i) Γ is a module homomorphism of degree $+1$ with respect to the $\mathbb{Z}$-gradation of the chain complex C and of C'.
(ii) Γ is a chain homotopy between φ and φ', that is, one has $\varphi' - \varphi = \Gamma d + d'\Gamma$.
(iii) We have the equalities $\alpha_{|P'_\star} = \gamma_{|P'_\star} = \alpha'_{|P'_\star}$.

We say that the two filtered chain morphisms (α, h) and (α', h') are *elementarily filtered chain homotopic*, and we write $(\alpha, h) \sim_e (\alpha', h')$, if there exists an elementary filtered chain homotopy between the maps (α, h) and (α', h'). We say that the two filtered chain morphisms (α, h) and (α', h') are *filtered chain homotopic*, and we write $(\alpha, h) \sim (\alpha', h')$, if there exists a sequence

$$(\alpha, h) = (\alpha_0, h_0) \sim_e (\alpha_1, h_1) \sim_e \dots \sim_e (\alpha_n, h_n) = (\alpha', h') \tag{4.17}$$

of filtered chain morphisms such that successive pairs are elementarily filtered chain homotopic.

The following proposition is straightforward.

Proposition 4.4.2 *The relation $\sim$ in the set of morphisms between the poset filtered chain complexes (P, C, d) and (P', C', d') in* PFCc *is an equivalence relation.* □

At first glance, it might not be immediately clear that there are morphisms that are filtered chain homotopic, but not elementarily filtered chain homotopic. We would therefore like to point out that Definition 4.4.1(iii) only requires that the maps α, α', and γ agree on the distinguished poset subset $P'_\star \subset P'$. Outside of this set, the maps can and generally will be different, and that can lead to morphisms that are filtered chain homotopic, but not elementarily filtered chain homotopic.

In certain situations, however, a filtered chain homotopy automatically happens to be an elementary filtered chain homotopy. To state a corresponding result, we say that a poset filtered chain complex (P, C, d) is *peeled* if we have $P_\star = P$. Then the following holds.

Proposition 4.4.3 *Assume that the poset filtered chain complex (P', C', d') is peeled. If $(\alpha, h), (\alpha', h') : (P, C, d) \to (P', C', d')$ are filtered chain homotopic, then $\alpha = \alpha'$, the domain of $\alpha = \alpha'$ is P', and we have in fact that $(\alpha, h) \sim_e (\alpha', h')$.*

Proof Choose $n + 1$ filtered morphisms $(\alpha_i, h_i) : (P, C, d) \to (P', C', d')$ as in (4.17) for $i = 0, \dots, n$. Let $(\gamma_i, \Gamma_i) : (P, C) \to (P', C')$ for $i \in \{1, 2, \dots, n\}$ be an elementary filtered chain homotopy between (α_{i-1}, h_{i-1}) and (α_i, h_i). Since we assumed $P'_\star = P'$, we have $\alpha_{i-1} = (\alpha_{i-1})_{|P'_\star} = (\gamma_i)_{|P'_\star} = \gamma_i$, as well as the identity $\alpha_i = (\alpha_i)_{|P'_\star} = (\gamma_i)_{|P'_\star} = \gamma_i$ for $i \in \{1, 2, \ldots, n\}$. It follows then that one has $\alpha_i = \alpha_{i-1}$ for all i, and thus $\alpha = \alpha'$. Moreover, we obtain

$$h_i - h_{i-1} = d'\Gamma_i + \Gamma_i d \tag{4.18}$$

for $i \in \{1, 2, \ldots, n\}$. Now define $\Gamma := \sum_{i=1}^{n} \Gamma_i$. Then one can easily verify that Γ is an α-filtered module homomorphism. Summing (4.18) for $i = 1, \ldots, n$, we get

$$h' - h = h_n - h_0 = d'\Gamma + \Gamma d.$$

Thus, the map (α, Γ) is an elementary filtered chain homotopy between the filtered morphisms (α, h) and (α', h'). □

4.5 Homotopy Category of Poset Filtered Chain Complexes

We refer to the equivalence classes of $\sim$ as the *homotopy equivalence classes*. They enable us to introduce the *homotopy category* of poset filtered chain complexes, denoted by CHPFCC, by taking poset filtered chain complexes as objects, homotopy equivalence classes of filtered chain morphisms in PFCC as morphisms in CHPFCC, and using the formula

$$[(\beta, g)]_\sim \circ [(\alpha, h)]_\sim := [(\alpha \circ \beta, g \circ h)]_\sim \tag{4.19}$$

for two arbitrarily given filtered morphisms $(\alpha, h) \in$ PFCC$((P, C, d), (P', C', d'))$ and $(\beta, g) \in$ PFCC$((P', C', d'), (P'', C'', d''))$ as the definition of composition of morphisms in CHPFCC. Finally, equivalence classes of identities in PFCC are identities in CHPFCC. The following result verifies that these definitions indeed lead to a category.

Proposition 4.5.1 *Assume that the filtered chain morphisms*

$$(\alpha, h), (\alpha', h') : (P, C, d) \to (P', C', d')$$

and

$$(\beta, g), (\beta', g') : (P', C', d') \to (P'', C'', d'')$$

are filtered chain homotopic, respectively. Then the compositions

$$(\beta, g) \circ (\alpha, h), (\beta', g') \circ (\alpha', h') : (P, C, d) \to (P'', C'', d'')$$

are also filtered chain homotopic. In particular, the category CHPFCC *is well-defined.*

Proof Assume that $(\alpha, h) \sim (\alpha', h')$ and $(\beta, g) \sim (\beta', g')$. Since $\sim$ is an equivalence relation, it suffices to prove that we have both $(\beta, g) \circ (\alpha, h) \sim (\beta, g) \circ (\alpha', h')$ and $(\beta, g) \circ (\alpha', h') \sim (\beta', g') \circ (\alpha', h')$. Using the same reasoning, we may assume that $(\alpha, h) \sim_e (\alpha', h')$ and $(\beta, g) \sim_e (\beta', g')$. Let $(\gamma, S) : (P, C) \to (P', C')$ be an elementary filtered chain homotopy between (α, h) and (α', h'). Consider the filtered morphism $(\eta, T) := (\beta, g) \circ (\gamma, S) : (P, C) \to (P'', C'')$. Then $\eta = \gamma\beta$, as well as $T = gS$. We will prove that (η, T) is an elementary filtered chain homotopy between $(\beta, g) \circ (\alpha, h)$ and $(\beta, g) \circ (\alpha', h')$. For this, we need to verify the three properties in Definition 4.4.1. Clearly, the map T is a $\mathbb{Z}$-graded module homomorphism of degree $+1$. Hence, property (i) is satisfied. To see (ii), observe that

$$gh' - gh = g(h' - h) = g(d'S + Sd) = d''gS + gSd = d''T + Td.$$

Finally, in view of the inclusion $\beta(P''_\star) \subset P'_\star$, which is a consequence of the definition of DPSET, one obtains

$$\alpha\beta_{|P''_\star} = \alpha_{|P'_\star}\beta_{|P''_\star} = \gamma_{|P'_\star}\beta_{|P''_\star} = \eta_{|P''_\star} = \gamma_{|P'_\star}\beta_{|P''_\star} = \alpha'_{|P'_\star}\beta_{|P''_\star} = \alpha'\beta_{|P''_\star}.$$

Altogether, we have verified that $(\beta, g) \circ (\alpha, h) \sim_e (\beta, g) \circ (\alpha', h')$, which in turn immediately implies that also $(\beta, g) \circ (\alpha, h) \sim (\beta, g) \circ (\alpha', h')$. The still missing equivalence $(\beta, g) \circ (\alpha', h') \sim (\beta', g') \circ (\alpha', h')$ can be shown similarly. □

After these preparations, the following notions are immediate. We say that a filtered chain morphism $(\alpha, h) : (P, C, d) \to (P', C', d')$ is a *filtered chain equivalence*, if its equivalence class $[(\alpha, h)]$ is an isomorphism in CHPFCC. In addition, we call two poset filtered chain complexes (P, C, d) and (P', C', d') *filtered chain homotopic*, if they are isomorphic in CHPFCC, i.e., if there exist two filtered chain morphisms $(\alpha, \varphi) : (P, C, d) \to (P', C', d')$ and $(\alpha', \varphi') : (P', C', d') \to (P, C, d)$ such that $(\alpha', \varphi') \circ (\alpha, \varphi)$ is filtered chain homotopic to $\mathrm{id}_{(P,C)}$, and $(\alpha, \varphi) \circ (\alpha', \varphi')$ is filtered chain homotopic to $\mathrm{id}_{(P',C')}$. In this situation, the filtered chain morphisms (α, φ) and (α', φ') are referred to as *mutually inverse* filtered chain equivalences. If the composition $(\alpha', \varphi') \circ (\alpha, \varphi)$ is elementarily filtered chain homotopic to $\mathrm{id}_{(P,C)}$, and $(\alpha, \varphi) \circ (\alpha', \varphi')$ is elementarily filtered chain homotopic to $\mathrm{id}_{(P',C')}$, we call the two morphisms (α, φ) and (α', φ') mutually inverse *elementary filtered chain equivalences*.

We finish this chapter with the following auxiliary proposition.

Proposition 4.5.2 *Assume that (P, C, d) is a poset filtered chain complex with P-gradation $(C_p)_{p \in P}$, and let $(W_p)_{p \in P}$ denote another P-gradation of C. If the gradations $(C_p)_{p \in P}$ and $(W_p)_{p \in P}$ are filtered equivalent in the sense of Definition 4.1.5, then the triple (P, W, d) with $W := \bigoplus_{p \in P} W_p$ and P-*

gradation $(W_p)_{p\in P}$ is also a poset filtered chain complex. Moreover, (P, W, d) and (P, C, d) are isomorphic in PFCc, *and therefore also in* ChPFCc.

Proof Note that both W and C are in fact the same modules, but the gradations $(W_p)_{p\in P}$ and $(C_p)_{p\in P}$ need not be the same. We know that d is a filtered homomorphism with respect to the $(C_p)_{p\in P}$ gradation of C. Since the $(W_p)_{p\in P}$ gradation of $W = C$ is filtered equivalent to the $(C_p)_{p\in P}$ gradation, we see from Proposition 4.1.6 that d is a filtered homomorphism with respect to the $(W_p)_{p\in P}$ gradation of C as well. Hence, (P, W, d) is a poset filtered chain complex. We will now prove that

$$(\mathrm{id}_P, \mathrm{id}_C) : (P, C, d) \to (P, W, d) \tag{4.20}$$

and

$$(\mathrm{id}_P, \mathrm{id}_W) : (P, W, d) \to (P, C, d) \tag{4.21}$$

are mutually inverse isomorphisms in PFCc. Let $p \in P$. Since both $(C_p)_{p\in P}$ and $(W_p)_{p\in P}$ are filtered equivalent, using Proposition 4.3.4(ii) we get

$$C_p \cong C_{p^\leq}/C_{p^<} = W_{p^\leq}/W_{p^<} \cong W_p.$$

Hence, the chain complex C_p is essential if and only if W_p is essential, and from this, it follows immediately that W satisfies (4.16), because C satisfies (4.16).

Since $(C_p)_{p\in P}$ and $(W_p)_{p\in P}$ are filtered equivalent, we get from Corollary 4.1.4 that $(\mathrm{id}_P, \mathrm{id}_C) : (P, C) \to (P, W)$ and $(\mathrm{id}_P, \mathrm{id}_W) : (P, W) \to (P, C)$ are filtered morphisms. Clearly, both are chain maps. It follows that the morphisms (4.20) and (4.21) are well-defined. Since they are also mutually inverse, the conclusion follows. □

Chapter 5
Algebraic Connection Matrices

After the preparations of the previous chapter, we are now in a position to introduce our notion of connection matrices in a purely algebraic way. While the definition is modeled on previous work by Robbin and Salamon [44], Spendlove [46], as well as Harker et al. [21], their approach has to be extended to allow for varying underlying posets. For this, we first introduce the notion of reduced filtered chain complexes, before discussing Conley complexes and connection matrices, as well as their existence for arbitrary poset filtered chain complexes. We close the chapter with a new equivalence relation for Conley complexes, which enables us to precisely formulate the uniqueness question of connection matrices for the first time. Despite being a completely algebraic criterion based on the notion of essentially graded morphisms, it will later allow us to detect underlying bifurcations in combinatorial dynamics.

5.1 Reduced Filtered Chain Complexes

Recall that by Corollary 4.3.6, every poset filtered chain complex (P, C, d) gives rise to the induced chain complexes (C_p, d_{pp}) for every $p \in P$. The following definition lies at the heart of the notion of Conley complexes.

Definition 5.1.1 (Reduced Filtered Chain Complex) We say that a poset filtered chain complex (P, C, d) is *reduced*, if it is peeled, i.e., if we have the equality $P_\star = P$, and if the chain complex (C_p, d_{pp}) is boundaryless for all $p \in P$.

Example 5.1.2 (Nonuniqueness via Subdivision, $\lhd$ 4.3.3 $\rhd$ 5.2.8) The filtered chain complex in Example 4.3.3 is not reduced, because one can check that d_{pp} is not boundaryless for $p = \mathbf{u}$. Consider therefore the different set of words

$$X' := \{\, \mathbf{A}, \mathbf{B}, \mathbf{AB}, \mathbf{AD}, \mathbf{AE} \,\},$$

M. Mrozek, T. Wanner, *Connection Matrices in Combinatorial Topological Dynamics*, SpringerBriefs in Mathematics,
https://doi.org/10.1007/978-3-031-87600-4_5

a subset $P^\sharp := \{\mathbf{p}, \mathbf{q}, \mathbf{r}, \mathbf{s}, \mathbf{t}\}$ of P, and a poset $\mathbb{P}^\sharp = (P^\sharp, \leq^\sharp)$ with its partial order $\leq^\sharp$ defined by the Hasse diagram

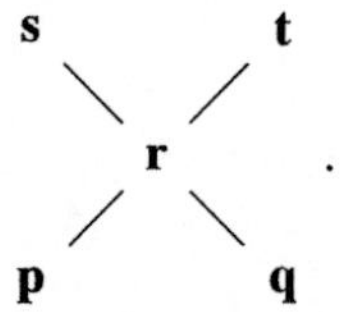

(5.1)

Notice that the partial order $\leq^\sharp$ is just the restriction to $P^\sharp$ of the partial order $\leq$ in the poset P defined in Example 4.3.3. Proceeding as in Example 4.3.3, we obtain a $\mathbb{P}^\sharp$-filtered $\mathbb{Z}_2$-module $(\mathbb{P}^\sharp, C', d')$ with $C'_p := \mathbb{Z}_2\langle X'_p\rangle$, where X'_p for $p = \mathbf{p}, \mathbf{q}, \mathbf{r}, \mathbf{s}, \mathbf{t}$ in $P^\sharp$ are defined, respectively, as

$$\{\mathbf{A}\}, \{\mathbf{B}\}, \{\mathbf{AB}\}, \{\mathbf{AD}\}, \{\mathbf{AE}\},$$

and the homomorphism $d' : C' \to C'$ is defined on the basis X' by the matrix

d'	**A**	**B**	**AB**	**AD**	**AE**
A			1	1	1
B			1		
AB					
AD					
AE					

. (5.2)

One can easily check that the partial order $\leq^\sharp$ is d'-admissible. However, it is not the native partial order of d'. As the reader may verify, the native partial order of d' is the partial order $\leq'$ given by the Hasse diagram

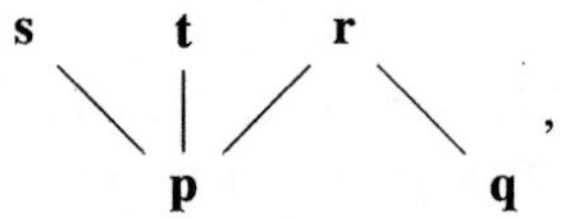

(5.3)

which gives a poset $\mathbb{P}' = (P^\sharp, \leq')$. Replacing the poset $\mathbb{P}^\sharp$ in $(\mathbb{P}^\sharp, C', d')$ by $\mathbb{P}'$, we obtain a $\mathbb{P}'$-filtered $\mathbb{Z}_2$-module $(\mathbb{P}', C', d')$ which differs from $(\mathbb{P}^\sharp, C', d')$ only in the d'-admissible partial order. Nevertheless, it is another object in PFCC. Note that both $(\mathbb{P}^\sharp, C', d')$ and $(\mathbb{P}', C', d')$ are peeled, because it follows from Proposition 3.5.9 that C'_p is homotopically essential for $p \in P^\sharp$. Therefore, we get from Proposition 4.4.3 that $(\mathbb{P}^\sharp, C', d')$ and $(\mathbb{P}', C', d')$ are neither isomorphic in PFCC nor in CHPFCC, because otherwise $\mathbb{P}^\sharp$ and $\mathbb{P}'$ would be isomorphic as posets. We note that both $(\mathbb{P}^\sharp, C', d')$ and $(\mathbb{P}', C', d')$ are also reduced, because $d'_{pp} = 0$ for every index $p \in P^\sharp$. ◊

Proposition 5.1.3 *Assume that (P, C, d) and (P', C', d') are two reduced poset filtered chain complexes and that $(\alpha, h), (\beta, g) : (P, C, d) \to (P', C', d')$ are filtered chain homotopic morphisms. If α is injective, then for $p \in P$ and $q \in P'$ we have*

$$\alpha(q) = p \Rightarrow h_{qp} = g_{qp}.$$

Proof Since (P', C', d') is reduced, it is peeled. Thus, we get from Proposition 4.4.3 that $\alpha = \beta : P' \to P$, and there exists an α-filtered degree $+1$ homomorphism $\Gamma : C \to C'$ such that $g - h = d'\Gamma + \Gamma d$. Since (P, C, d) and (P', C', d') are reduced, we get from Proposition 4.2.3 the identities

$$\begin{aligned} g_{qp} - h_{qp} &= (d'\Gamma)_{qp} + (\Gamma d)_{qp} = d'_{q\,\mathrm{id}(q)} \Gamma_{\mathrm{id}(q)p} + \Gamma_{q\alpha(q)} d_{\alpha(q)p} \\ &= 0\Gamma_{qp} + \Gamma_{qp} 0 = 0. \end{aligned}$$

For this, recall also that d and d' are id_P-filtered and that id_P is injective. This completes the proof. □

If two given poset filtered chain complexes (P, C, d) and (P', C', d') are filtered chain homotopic, then automatically the equivalence class of every filtered chain equivalence $(\alpha, \varphi) : (P, C, d) \to (P', C', d')$ is an isomorphism in CHPFCC. However, in the category PFCC, one has to prove that fact. This is addressed in the following result.

Theorem 5.1.4 *Suppose that (P, C, d) and (P', C', d') are two reduced poset filtered chain complexes. If they are filtered chain homotopic, then every filtered chain equivalence $(\alpha, \varphi) : (P, C, d) \to (P', C', d')$ is an isomorphism in* PFCC. *In particular, the poset filtered chain complexes (P, C, d) and (P', C', d') are isomorphic in* PFCC.

Proof Since (P, C, d) and (P', C', d') are reduced, we have $P_\star = P$ and $P'_\star = P'$. Let $(\alpha', \varphi') : (P', C', d') \to (P, C, d)$ be a filtered chain morphism such that one has both $(\alpha', \varphi') \circ (\alpha, \varphi) \sim \mathrm{id}_{(P,C,d)}$ and $(\alpha, \varphi) \circ (\alpha', \varphi') \sim \mathrm{id}_{(P',C',d')}$. Clearly, the maps id_P and $\mathrm{id}_{P'}$ are injective. Hence, it follows from Proposition 5.1.3 that

$$(\varphi'\varphi)_{pp} = (\mathrm{id}_C)_{pp} = \mathrm{id}_{C_p} \quad \text{and} \quad (\varphi\varphi')_{qq} = (\mathrm{id}_{C'})_{qq} = \mathrm{id}_{C'_q}$$

for arbitrary $p \in P$ and $q \in P'$. Therefore, one can deduce from Corollary 4.2.4 the equality $(\mathrm{id}_C)_{pp} = \varphi'_{p\alpha'(p)} \varphi_{\alpha'(p)p}$ for all $p \in P$. Similarly, we obtain for all $p' \in P'$ the identity $(\mathrm{id}_{C'})_{p'p'} = \varphi_{p'\alpha(p')} \varphi'_{\alpha(p')p'}$. Since α and α' are mutually inverse bijections, we may substitute $\alpha'(p)$ for p' to get $(\mathrm{id}_{C'})_{\alpha'(p)\alpha'(p)} = \varphi_{\alpha'(p)p} \varphi'_{p\alpha'(p)}$. From this, one can conclude that $\varphi_{\alpha'(p)p}$ and $\varphi'_{p\alpha'(p)}$ are mutually inverse module homomorphisms. Thus, it follows from Lemma 4.2.5 that (α, φ) is an isomorphism in FMOD. Since φ is a chain map, by Proposition 3.4.1, its inverse is also a chain map. Hence, the pair (α, φ) is an isomorphism in PFCc, which proves that (P, C, d) and (P', C', d') are isomorphic in PFCc. □

5.2 Conley Complexes and Connection Matrices

We now turn our attention to the definition of the Conley complex and the associated connection matrix. For this, we need two additional concepts.

Definition 5.2.1 (Representation of a Poset Filtered Chain Complex) Let (P, C, d) be an arbitrary poset filtered chain complex. By a *representation* of (P, C, d), we mean a triple $((P', C', d'), (\alpha, \varphi), (\beta, \psi))$ such that (P', C', d') is an object in PFCc, while $(\alpha, \varphi) : (P, C, d) \to (P', C', d')$ and $(\beta, \psi) : (P', C', d') \to (P, C, d)$ are mutually inverse elementary chain equivalences in which α and β are *strict*, that is, we have both $\operatorname{dom}\alpha = P'_\star$ and $\operatorname{dom}\beta = P_\star$. To simplify the terminology, in the sequel, we refer to the object (P', C', d') as a representation of (P, C, d), assuming that the associated mutually inverse elementary chain equivalences (α, φ) and (β, ψ) are implicitly given.

Proposition 5.2.2 *For any representation $((P', C', d'), (\alpha, \varphi), (\beta, \psi))$ of a poset filtered chain complex (P, C, d), the maps $\alpha : P'_\star \to P_\star$ and $\beta : P_\star \to P'_\star$ are mutually inverse order preserving bijections.*

Proof Since α and β are strict maps, we obtain from the definition of elementary chain homotopy both the identities $\mathrm{id}_{P_\star} = (\alpha\beta)_{P_\star} = \alpha_{P'_\star}\beta_{P_\star} = \alpha\beta$, as well as the equalities $\mathrm{id}_{P'_\star} = (\beta\alpha)_{P'_\star} = \beta_{P_\star}\alpha_{P'_\star} = \beta\alpha$. □

Definition 5.2.3 (Transfer Morphism) For an arbitrary poset filtered chain complex (P, C, d), we now consider the two representations $((P', C', d'), (\alpha, \varphi), (\beta, \psi))$ and $((P'', C'', d''), (\alpha', \varphi'), (\beta', \psi'))$. Then the *transfer morphism* from the poset filtered chain complex (P', C', d') to the poset filtered chain complex (P'', C'', d'') is defined as the filtered chain morphism $(\beta\alpha', \varphi'\psi)$.

We would like to point out that the pair of transfer morphisms $(\beta\alpha', \varphi'\psi)$ from (P', C', d') to (P'', C'', d'') and $(\beta'\alpha, \varphi\psi')$ from (P'', C'', d'') to (P', C', d') is mutually inverse chain equivalences.

After these preparations, we can now define the central concept of this book. For this, let (P, C, d) denote an arbitrary poset filtered chain complex. Then the following definition builds upon Definitions 5.1.1 and 5.2.1 and introduces both the notions of Conley complex and of connection matrix.

Definition 5.2.4 (Conley Complex and Connection Matrix) Let (P, C, d) denote an arbitrary poset filtered chain complex. Then a *Conley complex* of (P, C, d) is any reduced representation $(\bar{P}, \bar{C}, \bar{d})$ of (P, C, d). In addition, the $(\bar{P}, \bar{P})$-matrix of the boundary homomorphism $\bar{d}$ is called a *connection matrix* of the poset filtered chain complex (P, C, d). Finally, we would like to point out that without loss of generality one can always assume $\bar{P} = P_\star$, as the discussion in the next paragraph shows.

Since a Conley complex $(\bar{P}, \bar{C}, \bar{d})$ of a filtered chain complex (P, C, d) is reduced, we have $\bar{P}_\star = \bar{P}$, and, in view of Proposition 5.2.2, we see that $\bar{P}$ is order isomorphic to $P_\star$. Therefore, in a Conley complex $(\bar{P}, \bar{C}, \bar{d})$ of (P, C, d), we can identify the poset $\bar{P}$ with $P_\star$, the map α with the inclusion map $\iota_P : P_\star \hookrightarrow P$, and the map β with $\iota_P^{-1} : P \nrightarrow P_\star$. For the remainder of this book, we will make use of this specific representation.

Clearly, any Conley complex $(\bar{P}, \bar{C}, \bar{d})$ is chain homotopic to (P, C, d). Therefore, we have the following straightforward proposition.

Proposition 5.2.5 *If $(P_\star, \bar{C}, \bar{d})$ is any Conley complex of a poset filtered chain complex (P, C, d), then the homology module $H(\bar{C})$ is isomorphic to the homology module $H(C)$.* □

As an immediate consequence of Theorem 5.1.4, we obtain the following two corollaries, which show that the notion of Conley complex is well-defined.

Corollary 5.2.6 *The Conley complex of a poset filtered chain complex is uniquely determined up to an isomorphism in* PFCC. *In particular, the transfer morphism between two Conley complexes of a given poset filtered chain complex is an isomorphism in* PFCC. □

Corollary 5.2.7 *If two poset filtered chain complexes are filtered chain homotopic, that is, if they are isomorphic in* CHPFCC, *then their associated Conley complexes are isomorphic in* PFCC. □

Example 5.2.8 (Nonuniqueness via Subdivision, ◁ 5.1.2 ▷ 5.5.7) One can verify that the reduced poset filtered chain complex $(\mathbb{P}^\sharp, C', d')$ from Example 5.1.2, based on the poset $\mathbb{P}^\sharp = (P^\sharp, \leq^\sharp)$ with $\leq^\sharp$ given by the Hasse diagram (5.1), is a Conley complex of the poset filtered chain complex in Example 4.3.3. The associated connection matrix is given in (5.2). More precisely, consider the inclusion map $\varepsilon : P^\sharp \ni x \mapsto x \in P$ and the homomorphisms $h' : C \to C'$ and $g' : C' \to C$ given by the matrices

h'	A	B	AB	C	AC	BC	ABC	CD	CE
A	1			1					
B		1							
AB			1			1			
AD								1	
AE									1

and

g'	A	B	AB	AD	AE
A	1				
B		1			
AB			1		
C					
AC				1	1
BC					
ABC					
CD				1	
CE					1

.

One can verify that

$$(\varepsilon, h') : (P, C, d) \to (\mathbb{P}^\sharp, C', d') \text{ and } (\varepsilon^{-1}, g') : (\mathbb{P}^\sharp, C', d') \to (P, C, d),$$

where the partial map $\varepsilon^{-1} : P \nrightarrow P^\sharp$ is defined as the inverse relation of the map ε, are mutually inverse elementary filtered chain equivalences. This immediately implies that a Conley complex of the poset filtered chain complex (P, C, d) in Example 4.3.3 is given by the representation $((\mathbb{P}^\sharp, C', d'), (\varepsilon, h'), (\varepsilon^{-1}, g'))$. ♢

5.3 Existence of Conley Complexes

We now turn our attention to the question of existence, i.e., does every poset filtered chain complex have an associated Conley complex and connection matrix? Although in our setting the poset in a poset filtered chain complex is not fixed as in [21, 44, 46], the existence proof of a Conley complex for a poset filtered chain complex in our sense can be adapted from the argument in [44, Theorem 8.1, Corollary 8.2]. For the sake of completeness, we present the details. We begin with a technical lemma, which is a counterpart to [44, Theorem 8.1].

Lemma 5.3.1 *Assume that $P \neq \varnothing$ and that (P, C, d) denotes an arbitrary poset filtered chain complex with field coefficients. Then there exist four families $\{W_p\}_{p\in P}$, $\{V_p\}_{p\in P}$, $\{B_p\}_{p\in P}$, $\{H_p\}_{p\in P}$ of $\mathbb{Z}$-graded submodules of C such that the following statements are satisfied:*

(i) The family $\{W_p\}_{p\in P}$ is a P-gradation of C which is filtered equivalent to the gradation $\{C_p\}_{p\in P}$. In particular, $C = \bigoplus_{p\in P} W_p$.
(ii) For every $p \in P$, we have $W_p = V_p \oplus H_p \oplus B_p$.
(iii) The inclusion $d(V_p) \subset B_p$ holds, and $d_{|V_p} : V_p \to B_p$ is a module isomorphism, for every $p \in P$.
(iv) The identity $d_{pp}(H_p) = 0$ is satisfied for all $p \in P$.
(v) For $H := \bigoplus_{p\in P} H_p$, we have $d(H) \subset H$, and $(P, H, d_{|H})$ is a poset filtered chain complex.

Proof Assume that (P, C, d) is a poset filtered chain complex. We proceed by induction on $n :=$ card P. First assume that $n = 1$. Let p_* be the unique element of P. Then $C = C_{p_*}$. We set $W_{p_*} := C$. It follows that the P-gradations $(C_p)_{p\in P}$ and $(W_p)_{p\in P}$ are identical, and therefore they are filtered equivalent, i.e., (i) is satisfied. The existence of V_{p_*}, B_{p_*}, H_{p_*} satisfying properties (ii)–(v) then follows from Proposition 3.4.5.

Now assume that $n > 1$. Let $r \in P$ be a maximal element in P, and consider the down set $P' := P \setminus \{r\} \in \mathrm{Down}(P)$. Let $C' := \bigoplus_{p\in P'} C_p$. Since d is a filtered homomorphism, we have $d(C') \subset C'$. Thus, $(P', C', d_{|C'})$ is a P'-filtered chain complex. Since card $P' = n - 1$, by our induction hypothesis, there exist four indexed families $\{W_p\}_{p\in P'}$, $\{V_p\}_{p\in P'}$, $\{B_p\}_{p\in P'}$, $\{H_p\}_{p\in P'}$ of submodules such that properties (i)–(iv) hold for $(P', C', d_{|C'})$. In order to obtain respective families for the original complex (P, C, d), we will extend the families over P' to families over P by constructing in turn the modules V_r, B_r, H_r, and W_r.

To begin with, for a family $\{M_p\}_{p\in P'}$ of submodules of C which satisfies the identity $M_p \cap M_q = \{0\}$ for all $p \neq q$, and an $L \in \mathrm{Down}(P')$, we recall the notation

$$M_L := \bigoplus_{p\in L} M_p.$$

Set $C' := C_{P'}$, $W' := W_{P'}$, $V' := V_{P'}$, $B' := B_{P'}$, and $H' := H_{P'}$. Then from (i) and (ii) applied to $(P', C', d_{|C'})$, we get

$$C'_L = C_L = W_L = V_L \oplus H_L \oplus B_L \quad \text{for every} \quad L \in \mathrm{Down}(P'). \tag{5.4}$$

We will show that

$$V' \cap (H' + d(C)) = 0. \tag{5.5}$$

Indeed, if $x \in V'$ and $x = x_1 + dx_2$ for $x_1 \in H'$ and $x_2 \in C$, then one obtains $dx = dx_1$. By (v) of the induction assumption, $(P', H', d_{|H'})$ is a poset filtered chain complex. Hence, $dx = dx_1 = d_{|H'}x_1 \in H'$. Also, by (iii) of the induction assumption, we have the inclusion $dx \in B'$, and therefore $dx \in H' \cap B' = 0$. Since $x \in V'$, one then obtains $x = 0$. This proves (5.5).

Since $d^{-1}(C_{r^<}) \cap C_{r^\leq}$ is a $\mathbb{Z}$-graded submodule of $C_{r^\leq}$, we can find a $\mathbb{Z}$-graded submodule V_r of $C_{r^\leq}$ such that

$$C_{r^\leq} = (d^{-1}(C_{r^<}) \cap C_{r^\leq}) \oplus V_r, \tag{5.6}$$

where we also use the fact that the modules have field coefficients. We will prove

$$d^{-1}(C_{r^<}) \cap C_{r^\leq} = C_{r^<} + (d^{-1}(H_{r^<}) \cap C_{r^\leq}). \tag{5.7}$$

To see that the right-hand side of (5.7) is contained in the left-hand side, observe that the right-hand side is obviously contained in $C_{r^\leq}$. Since C is a filtered chain complex and $r^<$ is a down set, we have $d(C_{r^<}) \subset C_{r^<}$, as well as $C_{r^<} \subset d^{-1}(C_{r^<})$. We also have $d^{-1}(H_{r^<}) \subset d^{-1}(C_{r^<})$, because $H_{r^<} \subset W_{r^<}$ by (ii) and $W_{r^<} = C_{r^<}$ by (i). To prove the opposite inclusion, let $x \in C_{r^\leq} \cap d^{-1}(C_{r^<})$. Then $dx \in C_{r^<}$. Hence, by (5.4), we can find $x_V \in V_{r^<}$, $x_H \in H_{r^<}$, and $x_B \in B_{r^<}$ such that $dx = x_V + x_H + x_B$. From (iii), we get $x_B = dy_V$ for some $y_V \in V_{r^<}$. It follows that $x_V = d(x - y_V) - x_H$. Hence, $x_V \in V_{r^<} \cap (d(C) + H_{r^<}) \subset V' \cap (d(C) + H')$. Thus, from (5.5), one obtains the equality $x_V = 0$ and the inclusion $d(x - y_V) = x_H \in H_{r^<}$, which in turn implies the inclusion $x - y_V \in d^{-1}(H_{r^<})$. We also have $x \in C_{r^\leq}$ and $y_V \in V_{r^<} \subset C_{r^<} \subset C_{r^\leq}$. Therefore, one finally obtains $x = y_V + (x - y_V) \in C_{r^<} + (C_{r^\leq} \cap d^{-1}(H_{r^<}))$. This completes the proof of (5.7).

Now set $B_r := d(V_r)$. We will prove that

$$B_r \cap C_{r<} = 0. \tag{5.8}$$

Let $x \in B_r \cap C_{r<}$. Then $x = dy$ for some $y \in V_r \subset C_{r\leq}$. Since $x \in C_{r<}$, one obtains the inclusion $y \in d^{-1}(C_{r<})$. Therefore, $y \in V_r \cap C_{r\leq} \cap d^{-1}(C_{r<})$. It follows from (5.6) that $y = 0$. Hence, $x = d0 = 0$, which proves (5.8).

Since (P, C, d) is a poset filtered chain complex, the boundary homomorphism d is a filtered homomorphism. Thus, we have $C_{r<} \subset d^{-1}(C_{r<})$. Obviously, $C_{r<} \subset C_{r\leq}$ and $B_r = d(V_r) \subset d(C_{r\leq}) \cap d^{-1}(0) \subset C_{r\leq} \cap d^{-1}(H_{r<})$. Therefore, by (5.8), one has a direct sum of $\mathbb{Z}$-graded submodules

$$(C_{r<} \cap d^{-1}(H_{r<})) \oplus B_r \subset C_{r\leq} \cap d^{-1}(H_{r<}). \tag{5.9}$$

Hence, we can choose a $\mathbb{Z}$-graded submodule H_r such that

$$C_{r\leq} \cap d^{-1}(H_{r<}) = (C_{r<} \cap d^{-1}(H_{r<})) \oplus B_r \oplus H_r. \tag{5.10}$$

Thus, it follows from (5.7) and Proposition 3.3.1 that

$$C_{r\leq} \cap d^{-1}(C_{r<}) = C_{r<} \oplus B_r \oplus H_r. \tag{5.11}$$

Therefore, setting $W_r := V_r \oplus B_r \oplus H_r$, we get from (5.6) that

$$C_{r\leq} = C_{r<} \oplus W_r. \tag{5.12}$$

We now have well-defined families $\{W_p\}_{p\in P}$, $\{V_p\}_{p\in P}$, $\{B_p\}_{p\in P}$, $\{H_p\}_{p\in P}$ of submodules of C. We will prove that they indeed satisfy properties (i)–(v) for (P, C, d). To prove (i), observe that by (5.12)

$$C = C_{P'\cup r\leq} = C' + C_{r\leq} = C' + C_{r<} + W_r = C' + W_r.$$

We claim that $C = C' \oplus W_r$. Indeed, in view of identity (5.12), we have $W_r \subset C_{r\leq}$. Therefore, $W_r \cap C' \subset W_r \cap C_{r\leq} \cap C' = W_r \cap C_{r<} = 0$. This together with the induction assumption shows that

$$C = \bigoplus_{p\in P} W_p.$$

To show that $(W_p)_{p\in P}$ is filtered equivalent to $(C_p)_{p\in P}$, one needs to verify the equality $C_L = W_L$ for all $L \in \operatorname{Down}(P)$. Note that by the induction assumption

$$C_L = W_L \quad \text{for} \quad L \in \operatorname{Down}(P'). \tag{5.13}$$

Thus, we only need to consider the case when $r \in L$. For this, let $L' := L \setminus \{r\}$. Then one has $L = L' \cup r^{\leq}$, and (5.12) and (5.13) yield

$$C_L = C_{L'} + C_{r^{\leq}} = W_{L'} + C_{r^{<}} + W_r = W_{L'} + W_{r^{<}} + W_r = W_{L'} + W_{r^{\leq}} = W_L.$$

This proves property (i). By the induction assumption, properties (ii)–(v) need to be verified only for $p = r$. Property (ii) for $p = r$ follows from the definition of W_r. To see (iii) for $p = r$, take an $x \in V_r$ such that $dx = 0 \in C_{r^{<}}$. Since the inclusion $V_r \subset C_{r^{\leq}}$ holds, it follows that $x \in V_r \cap C_{r^{\leq}} \cap d^{-1}(C_{r^{<}})$, and we get from (5.6) that $x = 0$. Thus, $d_{|V_r}$ is a monomorphism. By the definition of B_r, it is an epimorphism, which proves (iii). Finally, by (5.10), we have $H_r \subset d^{-1}(H_{r^{<}})$, which implies

$$d(H_r) \subset H_{r^{<}}. \tag{5.14}$$

Therefore, $d_{rr}(H_r) = 0$, which proves (iv) for $p = r$. To show that $(P, H, d_{|H})$ is a poset filtered chain complex, we will prove that

$$d(H_L) \subset H_L \text{ for every } L \in \mathrm{Down}(P). \tag{5.15}$$

Property (5.15) holds by our induction assumption if one has $r \notin L$. Thus, we now assume $r \in L$ and set $L' := L \setminus \{r\}$. Then we have $H_L = H_{L'} + H_{r^{<}} + H_r$, and by (5.14) and the induction assumption, one further obtains

$$d(H_L) \subset d(H_{L'}) + d(H_{r^{<}}) + d(H_r) \subset H_{L'} + H_{r^{<}} \subset H_L,$$

which proves (5.15). Since $P \in \mathrm{Down}(P)$ and $H_P = H$, we get from (5.14) that the restriction $d_{|H} : H \to H$ is well-defined. Since $d^2 = 0$ and $d(H) \subset H$, we get $d_{|H}^2 = 0$. This proves that $(H, d_{|H})$ is a chain complex. From (5.15) and Corollary 4.1.4, one finally can conclude that $d_{|H}$ is a filtered homomorphism. This proves (v) and completes the proof of the lemma. □

After these preparations, we can now prove the main result of this chapter, which guarantees the existence of a Conley complex and associated connection matrix for every poset filtered chain complex with field coefficients.

Theorem 5.3.2 (Existence of the Conley Complex) *Every poset filtered chain complex with field coefficients admits a Conley complex and a connection matrix.*

Proof Assume that (P, C, d) is a poset filtered chain complex with field coefficients. Consider the four families of submodules $\{W_p\}_{p \in P}$, $\{V_p\}_{p \in P}$, $\{B_p\}_{p \in P}$, and $\{H_p\}_{p \in P}$ satisfying properties (i)–(v), as guaranteed by Lemma 5.3.1. Then, by Lemma 5.3.1(i), the collection $\{W_p\}_{p \in P}$ is a P-gradation of the module $W := \bigoplus_{p \in P} W_p$, which coincides with the module C, and, by Proposition 4.5.2, the triple (P, W, d) is a poset filtered chain complex which is filtered chain isomorphic to (P, C, d). Thus, it suffices to prove that (P, W, d) admits a Conley complex and a connection matrix, since the composition of a filtered chain isomorphism and an elementary filtered chain equivalence gives again an elementary filtered chain equivalence.

For this, set $V := \bigoplus_{p\in P} V_p$, $B := \bigoplus_{p\in P} B_p$, and $H := \bigoplus_{p\in P} H_p$. Then one has $W = V \oplus H \oplus B$. Moreover, (P, V), (P, B), and (P, H) are all objects of GMOD, and by Lemma 5.3.1(iv), the triple $(P, H, d_{|H})$ is a poset filtered chain complex. Clearly, it is a filtered chain subcomplex of (P, W, d).

Now define the sets $Q := Q_\star := P_\star$. Let $\alpha : Q \to P$ denote the inclusion map. Clearly, α is order preserving. Moreover, $\operatorname{dom}\alpha = Q = Q_\star$, and therefore α is strict. Since α is injective, the inverse relation $\beta := \alpha^{-1} : P \nrightarrow Q$ is a well-defined partial map which is also order preserving and strict, because $\operatorname{dom}\beta = \operatorname{im}\alpha = P_\star$. Hence, $\alpha : (Q, Q_\star) \to (P, P_\star)$ and $\beta : (P, P_\star) \to (Q, Q_\star)$ are well-defined strict morphisms in DPSET.

Recall that in view of Proposition 4.3.4(i), we have a chain complex (W_p, d_{pp}) for every $p \in P$, which has the homology decomposition $W_p = V_p \oplus H_p \oplus B_p$ due to Lemma 5.3.1. Thus, by Proposition 3.4.5(ii), the chain complexes (W_p, d_{pp}) and $(H_p, 0)$ are chain homotopic. It follows that (W_p, d_{pp}) is essential if and only if $(H_p, 0)$ is essential. We further know that $(H_p, 0)$ is essential if and only if $H_p \neq 0$, in view of Corollary 3.4.4. This establishes the inclusion

$$\{\, p \in P \mid H_p \neq 0 \,\} \subset P_\star. \tag{5.16}$$

We claim that $(Q, H, d_{|H})$ is a poset filtered chain complex. Since $Q = P_\star$, by (5.16), we have $H = \bigoplus_{p\in Q} H_p$. Let $L \in \operatorname{Down}(Q)$, and let $L' := \{\, p \in P \mid \exists_{q\in L}\ p \le q \,\}$. Then $L' \in \operatorname{Down}(P)$ and $H_p = 0$ for $p \in L' \setminus L$. Therefore, $H_{L'} = H_L$. Thus, since $d_{|H}$ is P-filtered, it follows from Corollary 4.1.4 that $d_{|H}$ is also Q-filtered, and $(Q, H, d_{|H})$ is indeed a poset filtered chain complex. Moreover, one obtains from the equalities $Q_\star = Q = P_\star$ and (5.16) that $(Q, H, d_{|H})$ satisfies (4.16). From Lemma 5.3.1(iv), we get that $(Q, H, d_{|H})$ is boundaryless. Since $Q = Q_\star$ by definition, we conclude that $(Q, H, d_{|H})$ is reduced.

Next, we will prove that $(Q, H, d_{|H})$ is filtered chain homotopic to the poset filtered chain complex (P, W, d). Let $\iota : H \to W$ and $\pi : W \to H$ denote the inclusion and projection homomorphisms, respectively. It follows from Lemma 5.3.1(v) that for $x \in H$, we have $d\iota x = dx = \iota dx$, that is, ι is a chain map. Also, for arbitrary $x \in W$, we get $x = x_V + x_H + x_B \in V \oplus H \oplus B$. Consequently, in view of Lemma 5.3.1(iii), one obtains the inclusion $dx = dx_V + dx_H + dx_B = dx_V + dx_H \in B \oplus H$. This in turn implies $\pi dx = dx_H = d\pi x$ and proves that π is a chain map.

The inclusion $\iota : H \to W$ is β-graded and thus also β-filtered. Similarly, the projection $\pi : W \to H$ is α-filtered. Therefore, we have two well-defined filtered morphisms $(\beta, \iota) : (Q, H, d_{|H}) \to (P, W, d)$ and $(\alpha, \pi) : (P, W, d) \to (Q, H, d_{|H})$.

Obviously, one has $(\alpha, \pi) \circ (\beta, \iota) = (\beta\alpha, \pi\iota) = (\operatorname{id}_Q, \operatorname{id}_H) = \operatorname{id}_{(Q,H)}$, which means that $(\alpha, \pi) \circ (\beta, \iota)$ is, trivially, elementarily filtered chain homotopic to $\mathrm{id}_{(Q,H)}$. We will now show that $(\beta, \iota) \circ (\alpha, \pi) = (\alpha\beta, \iota\pi)$ is elementarily filtered chain homotopic to $\mathrm{id}_{(P,W)}$. Let $\mu : W \ni x = x_V + x_H + x_B \mapsto x_B \in B$ denote the projection map, and let $\nu : V \ni x \mapsto x \in W$ denote the inclusion map. Clearly, both maps μ and ν are graded and, in consequence, filtered homomorphisms. By Lemma 5.3.1(iii), we have a well-defined P-graded degree -1 isomorphism $d_{|V} : V \ni x \mapsto dx \in B$ with a P-graded inverse, which is a degree $+1$ isomorphism $d_{|V}^{-1} : B \to V$. Taken altogether, the definition $\Gamma := \nu \circ d_{|V}^{-1} \circ \mu : W \to W$ gives a degree 1 filtered module homomorphism. We claim that

$$\mathrm{id}_C - \iota\pi = \Gamma d + d\Gamma. \tag{5.17}$$

To see (5.17), take an arbitrary $x \in W = C$. Then $x = x_V + x_H + x_B$, where we have $x_V \in V$, $x_H \in H$, and $x_B \in B$. Hence, one has $(\mathrm{id}_C - \iota\pi)(x) = x_V + x_B$, as well as $dx = dx_V + dx_H \in B + H$, $\Gamma dx = x_V$, $\Gamma x = d_{|V}^{-1}(x_B)$, and $d\Gamma x = x_B$. It follows that $(\Gamma d + d\Gamma)(x) = x_V + x_B$, which proves the identity (5.17). Clearly, the identity $(\beta\alpha)_{|P_\star} = \mathrm{id}_{P_\star} = \mathrm{id}_{P|P_\star}$ is satisfied. Therefore, the pair (id_P, Γ) is an elementary filtered chain homotopy between $(\alpha\beta, \iota\pi)$ and $\mathrm{id}_{(P,W)}$, which then implies that the poset filtered chain complexes $(Q, H, d_{|H})$ and (P, W, d) are filtered chain homotopic. Since we have already seen that $(Q, H, d_{|H})$ is reduced, it follows that the poset filtered chain complex $(Q, H, d_{|H})$ is in fact a Conley complex for the poset filtered chain complex (P, W, d)—and that the (Q, Q)-matrix of $d_{|H}$ is a connection matrix of (P, W, d). □

5.4 Conley Complexes of Subcomplexes

As we stated in Proposition 4.3.5, given a filtered chain complex (P, C, d), a convex subset $J \subset P$ induces a filtered chain complex (J, C_J, d_{JJ}). Hence, it is natural to wonder whether one can obtain a Conley complex of (J, C_J, d_{JJ}) from a Conley complex of (P, C, d). A positive answer to this question is given by the following theorem.

Theorem and Definition 5.4.1 *Assume that $(P_\star, \bar{C}, \bar{d})$ is a Conley complex of the poset filtered chain complex (P, C, d) and that $(\mathrm{id}_{P_\star}, \varphi) : (P, C, d) \to (P_\star, \bar{C}, \bar{d})$ and $(\mathrm{id}_{P_\star}, \psi) : (P_\star, \bar{C}, \bar{d}) \to (P, C, d)$ are the associated mutually inverse elementary chain equivalences. Let $J \subset P_\star$ be a convex subset and let $\hat{J} :=$ $\mathrm{conv}_P(J)$ denote the convex hull of J in P. Consider the restrictions*

$$\varphi_| := \varphi_{J\hat{J}} \quad \textit{and} \quad \psi_| := \psi_{\hat{J}J}.$$

Then $(J, \bar{C}_J, \bar{d}_{JJ})$ is a Conley complex of $(\hat{J}, C_{\hat{J}}, d_{\hat{J}\hat{J}})$, with $(\mathrm{id}_J, \varphi_|)$ and $(\mathrm{id}_J, \psi_|)$ as the associated mutually inverse elementary chain equivalences. We will call it the restriction *of the Conley complex $(P_\star, \bar{C}, \bar{d})$ to J.*

Proof We will first show that $(\mathrm{id}_J, \psi_|)$ is a morphism in PFCC. To prove that $\psi_|$ is id_J-filtered, assume that $(\psi_|)_{pq} \neq 0$ for some $p \in \hat{J}$ and $q \in J$. Since $(\psi_|)_{pq} = \psi_{pq}$ and ψ is $\mathrm{id}_{P_\star}$-filtered, we see that $p \leq p_1$ for some $p_1 \in P_\star$ such that $p_1 \leq q$. Hence, we get $p_1 \in \hat{J}$ by (3.1), because $p \leq p_1 \leq q$ and $p, q \in J$.

To complete the proof that $(\mathrm{id}_J, \psi_|)$ is a morphism in PFCC, we still need to verify that $\psi_| : (\bar{C}_J, \bar{d}_{JJ}) \to (C_{\hat{J}}, d_{\hat{J}\hat{J}})$ is a chain map. Observe that $\psi : (\bar{C}, \bar{d}) \to (C, d)$, as a chain map, satisfies the identity $\psi\bar{d} = d\psi$. Thus, in order to prove that $\psi_|$ is a chain map, it suffices to verify the following two equalities

$$(\psi\bar{d})_{\hat{J}J} = \psi_{\hat{J}J}\bar{d}_{JJ}, \tag{5.18}$$

$$(d\psi)_{\hat{J}J} = d_{\hat{J}\hat{J}}\psi_{\hat{J}J}. \tag{5.19}$$

Since for $p \in P$, $q \in P_\star$, we have

$$(\psi\bar{d})_{pq} = \sum_{r \in P_\star} \psi_{pr}\bar{d}_{rq},$$

in order to prove (5.18), it suffices to verify that

$$\psi_{pr}\bar{d}_{rq} \neq 0, \;\; p \in \hat{J}, \; q \in J \;\Rightarrow\; r \in J. \tag{5.20}$$

Hence, assume that $\psi_{pr}\bar{d}_{rq} \neq 0$ for some $p \in \hat{J}$ and $q \in J$. Then $\psi_{pr} \neq 0$, $\bar{d}_{rq} \neq 0$, and, using the fact that ψ is $\mathrm{id}_{P_\star}$-filtered, we can find a $p_1 \in P_\star$ such that $p \leq p_1$ and $p_1 \leq r$. In view of $p \in \hat{J} = \mathrm{conv}_p(J)$, by (3.1), we can also find a $p_- \in J \subset P_\star$ such that $p_- \leq p \leq p_1$ and, in consequence, $p_- \leq p_1 \leq r$. Since $\bar{d}_{rq} \neq 0$ and $\bar{d}$ is filtered, we also have $r \leq q$. But, $q \in J$, as well as $p_- \in J$. Now, again by (3.1), one obtains from $p_- \leq r \leq q$ the inclusion $r \in J$. This proves (5.20) and (5.18).

Similarly, to prove (5.19), it suffices to verify that

$$d_{pr}\psi_{rq} \neq 0, \;\; p \in \hat{J}, \; q \in J \;\Rightarrow\; r \in \hat{J}. \tag{5.21}$$

Hence, assume that $d_{pr}\psi_{rq} \neq 0$ for some $p \in \hat{J}$ and $q \in J$. Then $\psi_{rq} \neq 0$, $d_{pr} \neq 0$, and, using the fact that ψ is $\mathrm{id}_{P_\star}$-filtered, we can find an $r_1 \in P_\star$ such that $r \leq r_1$ and $r_1 \leq q$. From $d_{pr} \neq 0$ and the fact that d is filtered, one obtains $p \leq r$. In view of $p \in \hat{J} = \mathrm{conv}_P(J)$, we can find an element $p_- \in J$ such that $p_- \leq p \leq r$. Hence, one obtains $p_- \leq r \leq r_1 \leq q$. Since $p_-, q \in J$, we get $r \in \hat{J}$, which verifies (5.21) and (5.19). This completes the proof that the map $\psi_{\hat{J}J}$ is indeed a chain map and a morphism in PFCC.

By an analogous argument, we will now prove that

$$\varphi_| : (C_{\hat{J}}, d_{\hat{J}\hat{J}}) \to (\bar{C}_J, \bar{d}_{JJ})$$

is a morphism in PFCC. First, assume that $(\varphi_|)_{pq} \neq 0$ for some $p \in J$ and $q \in \hat{J}$. Then $p \leq q$, which shows that $\varphi_|$ is id_J-filtered. To see that $\varphi_|$ is a chain map, it suffices to verify the following two equalities

$$(\varphi d)_{J\hat{J}} = \varphi_{J\hat{J}}\bar{d}_{\hat{J}\hat{J}}, \tag{5.22}$$

$$(\bar{d}\varphi)_{J\hat{J}} = \bar{d}_{JJ}\varphi_{J\hat{J}}, \tag{5.23}$$

which reduces to verifying that

$$\varphi_{pr}d_{rq} \neq 0,\ p \in J,\ q \in \hat{J}\ \Rightarrow\ r \in \hat{J}, \tag{5.24}$$

$$\bar{d}_{pr}\varphi_{rq} \neq 0,\ p \in J,\ q \in \hat{J}\ \Rightarrow\ r \in J. \tag{5.25}$$

In order to establish the implication (5.24), assume that $\varphi_{pr}d_{rq} \neq 0$ for some $p \in J$ and $q \in \hat{J}$. Then $r \leq q$, and also $p \leq r$, because $p \in J \subset P_\star$. Hence, one obtains from (3.1) that $r \in \hat{J}$, which proves both (5.24) and (5.22). For the verification of the implication (5.25), assume that $\bar{d}_{pr}\varphi_{rq} \neq 0$ for some $p \in J$ and $q \in \hat{J}$. Then $p \leq r$, and also $r \leq q$, because $r \in P_\star$. Since $q \in \hat{J}$, by (3.1), we can choose a $q^+ \in J$ such that $q \leq q^+$, and therefore $r \leq q^+ \in J$. We also have $p \leq r$, because $\bar{d}$ is filtered. Thus, since J is convex, one obtains $r \in J$, which proves both (5.25) and (5.23). This completes the proof that also $\varphi_{J\hat{J}}$ is a chain map and a morphism in PFCC.

We still need to verify that the pairs $(\mathrm{id}_J, \varphi_|)$ and $(\mathrm{id}_J, \psi_|)$ are mutually inverse elementary chain equivalences. Since $(\mathrm{id}_{P_\star}, \varphi)$ and $(\mathrm{id}_{P_\star}, \psi)$ are in fact mutually inverse elementary chain equivalences, there exist two filtered, degree $+1$ homomorphisms $\Gamma : C \to C$ and $\Gamma' : \bar{C} \to \bar{C}$ such that

$$\psi\varphi = \mathrm{id}_C + \Gamma d + d\Gamma,$$

$$\varphi\psi = \mathrm{id}_{\bar{C}} + \Gamma'\bar{d} + \bar{d}\Gamma'.$$

Then we also have the identities

$$(\psi\varphi)_{\hat{J}\hat{J}} = \mathrm{id}_{C_{\hat{J}}} + (\Gamma d)_{\hat{J}\hat{J}} + (d\Gamma)_{\hat{J}\hat{J}}, \tag{5.26}$$

$$(\varphi\psi)_{JJ} = \mathrm{id}_{\bar{C}_J} + (\Gamma'\bar{d})_{JJ} + (\bar{d}\Gamma')_{JJ}. \tag{5.27}$$

Analogously to (5.18)–(5.19) and 5.22–(5.23), one can further verify that

$$(\psi\varphi)_{\hat{J}\hat{J}} = \psi_{\hat{J}J}\varphi_{J\hat{J}},$$

$$(\Gamma d)_{\hat{J}\hat{J}} = \Gamma_{\hat{J}\hat{J}}d_{\hat{J}\hat{J}},$$

$$(d\Gamma)_{\hat{J}\hat{J}} = d_{\hat{J}\hat{J}}\Gamma_{\hat{J}\hat{J}},$$

$$(\varphi\psi)_{JJ} = \varphi_{J\hat{J}}\psi_{\hat{J}J},$$

$$(\Gamma'\bar{d})_{JJ} = \Gamma'_{JJ}\bar{d}_{JJ},$$
$$(d\Gamma)_{JJ} = \bar{d}_{JJ}\Gamma'_{JJ}.$$

Applying these formulas to (5.26)–(5.27), we finally obtain the equations

$$\psi_{|}\varphi_{|} = \mathrm{id}_{C_{\hat{J}}} + \Gamma_{\hat{J}\hat{J}}d_{\hat{J}\hat{J}} + d_{\hat{J}\hat{J}}\Gamma_{\hat{J}\hat{J}},$$
$$\varphi_{|}\psi_{|} = \mathrm{id}_{\bar{C}_J} + \Gamma'_{JJ}\bar{d}_{JJ} + \bar{d}_{JJ}\Gamma'_{JJ},$$

which completes the proof. □

As an immediate consequence of Theorem 5.4.1 and Proposition 5.2.5, we note the following corollary.

Corollary 5.4.2 *In the situation of Theorem 5.4.1, for every convex subset $J \subset P_\star$, the homology module $H(C_{\hat{J}})$ is isomorphic to the homology module $H(\bar{C}_J)$.* □

5.5 Equivalence of Conley Complexes

The last three sections demonstrated that every poset filtered chain complex (P, C, d) does have an associated Conley complex, which is unique up to an isomorphism in PFCC. Moreover, while different representations of the Conley complex might lead to different connection matrices, any two of them are similar to each other via an isomorphism in the category PFCC. However, in the application of the classical connection matrix theory to dynamical systems, it was already pointed out by Franzosa [20] and Reineck [41, 42] that the connection matrix of a given Morse decomposition may not be unique in dynamic terms, which reflects underlying bifurcations in the dynamical system.

This dynamics-related lack of uniqueness of connection matrices can also be described directly in the algebraic setting of the present chapter, if one considers a stronger equivalence relation between Conley complexes. For this, we first have to discuss the notion of essentially graded morphisms, which can be defined as follows.

Definition 5.5.1 (Essentially Graded Morphism) Consider two poset filtered chain complexes (P, C, d) and (P', C', d'), and let $(\alpha, h) : (P, C, d) \to (P', C', d')$ be a filtered chain morphism. Then (α, h) is called *essentially graded*, if there exists a graded chain morphism $(\beta, g) : (P, C, d) \to (P', C', d')$ which is filtered chain homotopic to (α, h). In this case, the morphism (β, g) is called a *graded representative* of the filtered chain morphism (α, h).

Essentially graded morphisms between reduced poset filtered chain complexes are of particular interest in the following, since of course the Conley complex is always reduced. As the following result demonstrates, such morphisms have a unique graded chain morphism in their homotopy equivalence class.

Proposition and Definition 5.5.2 (Graded Representation) *Consider an essentially graded chain equivalence $(\alpha, h) : (P, C, d) \to (P', C', d')$, and assume that the poset filtered chain complexes (P, C, d) and (P', C', d') are reduced. Then there exists a unique graded chain morphism $(\beta, g) : (P, C, d) \to (P', C', d')$ in the homotopy equivalence class $[(\alpha, h)]$. We call it the* graded representation *of (α, h).*

Proof The existence of (β, g) follows from the definition of an essentially graded morphism. To prove uniqueness, assume that (β, g) and (β', g') are two graded chain morphisms in the homotopy equivalence class $[(\alpha, h)]$. Since (α, h) is a chain equivalence, there is a filtered chain morphism $(\alpha', h') : (P', C', d') \to (P, C, d)$ such that $(\alpha\alpha', h'h) \sim \mathrm{id}_{(P,C,d)}$ and $(\alpha'\alpha, hh') \sim \mathrm{id}_{(P',C',d')}$. Hence, it follows from Proposition 4.4.3 that $\alpha\alpha' = \mathrm{id}_P$ and $\alpha'\alpha = \mathrm{id}_{P'}$. In particular, the map α is injective. Since $(\beta, g) \sim (\alpha, h)$ and $(\beta', g') \sim (\alpha, h)$, we see that $(\beta, g) \sim (\beta', g')$, and from Proposition 4.4.3, one further obtains $\beta = \alpha = \beta'$. Hence, the map β is injective. Consider now $p \in P$ and $q \in P'$. If one assumes $p = \alpha(q)$, then we have the equality $g_{qp} = g'_{qp}$ from Proposition 5.1.3. If, on the other hand, one has $p \neq \alpha(q)$, then we get $g_{qp} = 0 = g'_{qp}$ from (4.3), because both g and g' are α-graded. □

The following proposition is an immediate consequence of Proposition 4.5.1 and the fact that the composition of graded chain morphisms is again a graded chain morphism. The latter statement in turn follows from the fact that the composition of graded morphisms is a graded morphism and that the composition of chain maps is again a chain map. Thus, we have the following result.

Proposition 5.5.3 *Assume that the two morphisms $(\alpha, h) : (P, C, d) \to (P', C', d')$ and $(\alpha', h') : (P', C', d') \to (P'', C'', d'')$ are filtered chain morphisms which are essentially graded, and suppose further that they have graded representatives given by $(\beta, g) : (P, C, d) \to (P', C', d')$ and $(\beta', g') : (P', C', d') \to (P'', C'', d'')$, respectively. Then the composition $(\alpha', h') \circ (\alpha, h)$ is also essentially graded and has the graded representative $(\beta', g') \circ (\beta, g)$.* □

It is straightforward to observe that every identity morphism in PFCC is essentially graded and that it is its own graded representation. Thus, in view of Proposition 5.5.3, we have a well-defined wide subcategory EGPFCC of PFCC whose morphisms are essentially graded morphisms in PFCC.

Proposition 5.5.4 *Assume that both (P, C, d) and (P', C', d') are reduced poset filtered chain complexes and that the two pairs $(\alpha, h) : (P, C, d) \to (P', C', d')$ and $(\alpha', h') : (P', C', d') \to (P, C, d)$ are mutually inverse chain equivalences. Suppose further that (α, h) is essentially graded and has the graded representation $(\bar{\alpha}, \bar{h})$. Then (α', h') is essentially graded as well, the pair $(\bar{\alpha}, \bar{h})$ is an isomorphism in* PGCC, *and $(\bar{\alpha}, \bar{h})^{-1}$ is the graded representation of (α', h').*

Proof By Definition 5.5.1, we have $(\alpha, h) \sim (\bar{\alpha}, \bar{h})$. Hence, one obtains from Proposition 4.5.1 the equivalences

$$(\bar{\alpha}, \bar{h}) \circ (\alpha', h') \sim (\alpha, h) \circ (\alpha', h') \sim \mathrm{id}_{(P', C', d')}, \tag{5.28}$$
$$(\alpha', h') \circ (\bar{\alpha}, \bar{h}) \sim (\alpha', h') \circ (\alpha, h) \sim \mathrm{id}_{(P, C, d)},$$

which imply that $(\bar{\alpha}, \bar{h})$ is a filtered chain equivalence. Now Theorem 5.1.4 shows that $(\bar{\alpha}, \bar{h})$ is an isomorphism in PFCC, and it follows that $(\bar{\alpha}, \bar{h})$ is an isomorphism in FMOD. Thus, we get from Proposition 4.4.3 that $\bar{\alpha} = \alpha$, and from Lemma 4.2.5 that α is an order isomorphism. In consequence, the inverse of $(\bar{\alpha}, \bar{h}) = (\alpha, \bar{h})$ in PFCC takes the form $(\alpha^{-1}, \bar{h}^{-1})$, and, clearly, the inverse $\bar{h}^{-1}$ is α^{-1}-graded. Moreover, from the equivalences in (5.28), we obtain

$$(\alpha', h') = (\alpha^{-1}, \bar{h}^{-1}) \circ (\alpha, \bar{h}) \circ (\alpha', h') \sim (\alpha^{-1}, \bar{h}^{-1}),$$

which proves that (α', h') is essentially graded with $(\alpha, \bar{h})^{-1} = (\bar{\alpha}, \bar{h})^{-1}$ as its graded representation. □

After this algebraic intermission, we return to our study of Conley complexes. As mentioned earlier, dynamical considerations require a stronger notion of equivalence of Conley complexes than the one provided by isomorphisms in PFCC. This equivalence is based on essentially graded morphisms and can be defined as follows.

Definition 5.5.5 (Equivalent Conley Complexes) We say that two Conley complexes of a poset filtered chain complex are *equivalent* if the associated transfer homomorphism is essentially graded.

Note that the transfer homomorphism is always an isomorphism in PFCC. For equivalence, one needs in addition that it is essentially graded, i.e., that it is filtered chain homotopic to a graded chain isomorphism.

It follows from Proposition 5.5.3 that the equivalence of Conley complexes of a given poset filtered chain complex is indeed an equivalence relation. Two equivalent Conley complexes, as well as the associated connection matrices, are basically the same. This justifies the following definition.

Definition 5.5.6 (Uniqueness of Conley Complexes and Connection Matrices) A poset filtered chain complex has a *uniquely determined Conley complex and connection matrix* if any two of its Conley complexes are equivalent.

Note that two nonequivalent Conley complexes of a poset filtered chain complex may be graded-conjugate, and the associated connection matrices may be graded similar. Thus, graded similarity of connection matrices is not sufficient for their uniqueness. It is indeed necessary to consider the specific transfer homomorphisms. This is illustrated in the following example.

Example 5.5.7 (Nonuniqueness via Subdivision, $\lhd$ 5.2.8 $\rhd$ 7.1.2) The filtered chain complex in Example 4.3.3 does not have a uniquely determined Conley complex. To see this, consider the set of words

$$X'' := \{\, \mathbf{A}, \mathbf{B}, \mathbf{AB}, \mathbf{BD}, \mathbf{BE} \,\}.$$

Similar to Example 4.3.3, we obtain a reduced, $\mathbb{P}^\sharp$-filtered $\mathbb{Z}_2$-module $(\mathbb{P}^\sharp, C', d')$ with $C''_p := \mathbb{Z}_2\langle X''_p\rangle$, where X''_p for $p = \mathbf{p}, \mathbf{q}, \mathbf{r}, \mathbf{s}, \mathbf{t} \in P^\sharp$ is defined respectively as

$$\{\mathbf{A}\}, \{\mathbf{B}\}, \{\mathbf{AB}\}, \{\mathbf{BD}\}, \{\mathbf{BE}\},$$

and the homomorphism $d'' : C'' \to C''$ is defined on the basis X'' by the matrix

d''	**A**	**B**	**AB**	**BD**	**BE**
A			1		
B			1	1	1
AB					
BD					
BE					

. (5.29)

Analogously to Example 5.1.2, one can immediately verify that the partial order $\leq^\sharp$ in $\mathbb{P}^\sharp$ is d''-admissible, but it is not the native partial order of d''. This native partial order of d'' is $\leq''$, defined by the Hasse diagram

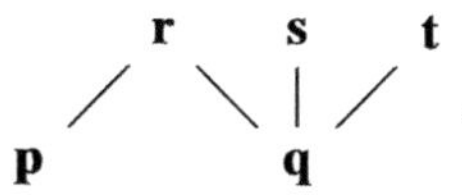

, (5.30)

and gives the poset $\mathbb{P}'' = (P^\sharp, \leq'')$. Then $(\mathbb{P}'', C'', d'')$ is a well-defined $\mathbb{P}''$-filtered, reduced chain complex which differs from $(\mathbb{P}^\sharp, C'', d'')$ only in the partial order. However, our interest in $(\mathbb{P}^\sharp, C'', d'')$ comes from the fact that it provides another Conley complex of the poset filtered chain complex in Example 4.3.3. For this, we consider again the inclusion map $\varepsilon : P^\sharp \ni x \mapsto x \in P$, as well as the two homomorphisms $h'' : C \to C''$ and $g'' : C'' \to C$ given by the matrices

h''	**A**	**B**	**AB**	**C**	**AC**	**BC**	**ABC**	**CD**	**CE**
A	1								
B		1		1					
AB			1		1				
BD								1	
BE									1

and

g''	**A**	**B**	**AB**	**BD**	**BE**
A	1				
B		1			
AB			1		
C					
AC					
BC				1	1
ABC					
CD				1	
CE					1

.

Then one can verify that $((\mathbb{P}^\sharp, C'', d''), (\varepsilon, h''), (\varepsilon^{-1}, g''))$ is another reduced representation, that is, a Conley complex of the poset filtered chain complex (P, C, d) presented in Example 4.3.3.

We claim that this Conley complex is not equivalent to the Conley complex $(\mathbb{P}^\sharp, C', d')$ presented in Example 5.2.8. To see this, assume to the contrary that the transfer homomorphism from $(\mathbb{P}^\sharp, C', d')$ to $(\mathbb{P}^\sharp, C'', d'')$ is essentially graded. Observe that this transfer homomorphism is $(\varepsilon^{-1}\varepsilon, h''g') = (\mathrm{id}_{P^\sharp}, h''g')$, and $h''g'$ has the matrix

$h''g'$	**A**	**B**	**AB**	**AD**	**AE**
A	1				
B		1			
AB			1	1	1
BD				1	
BE					1

.

Moreover, let (γ, f) denote a graded representation of $(\mathrm{id}_{P^\sharp}, h''g')$. Then we clearly have $(\gamma, f) \sim (\mathrm{id}_{P^\sharp}, h''g')$, and from Proposition 4.4.3, one obtains $\gamma = \mathrm{id}_{P^\sharp}$ and

$$h''g' - f = d''\Gamma + \Gamma d' \tag{5.31}$$

for some $\mathrm{id}_{P^\sharp}$-filtered degree $+1$ homomorphism $\Gamma : C' \to C''$. If we now evaluate both sides of (5.31) at the element **AD**, then one obtains

$$\mathbf{AB} + \mathbf{BD} - f(\mathbf{AD}) = (d''\Gamma)(\mathbf{AD}) + \Gamma(\mathbf{A}) = \Gamma(\mathbf{A}), \tag{5.32}$$

because $C''_2 = 0$. Since f is $\mathrm{id}_{P^\sharp}$-graded, we also get $\langle f(\mathbf{AD}), \mathbf{AB}\rangle = 0$. Hence, from (5.32), one can deduce the identity

$$1 = \langle \mathbf{AB} + \mathbf{BD} - f(\mathbf{AD}), \mathbf{AB}\rangle = \langle \Gamma(\mathbf{A}), \mathbf{AB}\rangle.$$

It follows that $\Gamma_{\mathbf{r},\mathbf{p}} \neq 0$, and, since Γ is $\mathrm{id}_{P^\sharp}$-filtered, one immediately obtains the inequality $\mathbf{r} \leq \mathbf{p}$. However, we see from the Hasse diagram (5.1) of the poset $\mathbb{P}^\sharp$ that $\mathbf{p} < \mathbf{r}$. This contradiction proves that the two Conley complexes $(\mathbb{P}^\sharp, C', d')$ and $(\mathbb{P}^\sharp, C'', d'')$ of (P, C, d) are not equivalent. The associated connection matrices are the matrices of d' and d'', which are shown in (5.2) and (5.29), respectively. Note that despite the fact that these matrices are not equivalent as connection matrices of (P, C, d), they are graded similar. To see this, define the map $\chi : \mathbb{P}^\sharp \to \mathbb{P}^\sharp$ by

$$\chi(x) := \begin{cases} \mathbf{q} & \text{if } x = \mathbf{p}, \\ \mathbf{p} & \text{if } x = \mathbf{q}, \\ x & \text{otherwise,} \end{cases}$$

and $h : C' \to C''$ by the matrix

h	**A**	**B**	**AB**	**AD**	**AE**
A		1			
B	1				
AB			1		
BD				1	
BE					1

.

Then one can easily verify that χ is an order preserving bijection, that (χ, h) is a graded chain isomorphism, and that we have

$$(\mathrm{id}_{P^\sharp}, d'') \circ (\chi, h) = (\chi, h) \circ (\mathrm{id}_{P^\sharp}, d').$$

This readily establishes their graded similarity. ◊

Chapter 6
Connection Matrices in Lefschetz Complexes

In this chapter, we turn our attention to a more specialized situation. Rather than continuing to study connection matrices for general poset filtered chain complexes, we now consider the setting of Lefschetz complexes. Their definition and basic properties have already been recalled in Sect. 3.5, and therefore this chapter focuses on the introduction of acyclic partitions in Lefschetz complexes, as well as their associated connection matrices. We then concentrate on the fundamental properties of a special case, the so-called singleton partition, and end the chapter with a discussion of refinements.

6.1 Connection Matrices of Acyclic Partitions

In order to define connection matrices on arbitrary Lefschetz complexes, as introduced in Definition 3.5.1, we make use of the following fundamental concept.

Definition 6.1.1 (Acyclic Partition) Let (X, κ) be a Lefschetz complex equipped with the Lefschetz topology induced by the face relation partial order $\leq_\kappa$. Furthermore, let $\mathcal{E} = (E_p)_{p \in P}$ be a partition of X into locally closed sets in this Lefschetz topology. Consider the relation $\preceq_\mathcal{E}$ in P defined for $p, q \in P$ by

$$p \preceq_\mathcal{E} q \quad \text{if and only if} \quad E_p \cap \operatorname{cl} E_q \neq \varnothing. \tag{6.1}$$

The collection $\mathcal{E}$ is called an *acyclic partition* of X if $\preceq_\mathcal{E}$ can be extended to a partial order on P. Every such extension is called $\mathcal{E}$-admissible. Note that then the smallest $\mathcal{E}$-admissible partial order, that is, the intersection of all the $\mathcal{E}$-admissible extensions of $\preceq_\mathcal{E}$, equals the transitive closure of $\preceq_\mathcal{E}$. We call it the *inherent* partial order of $\mathcal{E}$ and denote it by the symbol $\leq_\mathcal{E}$.

M. Mrozek, T. Wanner, *Connection Matrices in Combinatorial Topological Dynamics*, SpringerBriefs in Mathematics,
https://doi.org/10.1007/978-3-031-87600-4_6

We recall (see Sect. 3.1) that given an arbitrary subset $I \subset P$ we use the compact notation $|I| := \bigcup_{p\in I} E_p \subset X$ to denote the subset of X associated with I. The following proposition characterizes some topological features of $|I|$ in terms of poset features of I.

Proposition 6.1.2 *Assume that $\mathcal{E} = (E_p)_{p\in P}$ is an acyclic partition of a Lefschetz complex X, that $\leq$ is an $\mathcal{E}$-admissible partial order, and that $I \subset P$. Then the following statements hold:*

(i) If I is a down set with respect to $\leq$, then $|I|$ is closed.
(ii) If I is convex with respect to $\leq$, then $|I|$ is locally closed.

Proof In order to prove (i), we assume that $I \subset P$ is a down set and let $x \in \operatorname{cl}|I|$. Then there exists a $q \in I$ such that $x \in \operatorname{cl} E_q$. Since $\mathcal{E}$ is a partition of X, there exists a $p \in P$ such that $x \in E_p$. It follows that $E_p \cap \operatorname{cl} E_q \neq \varnothing$, and since $\leq$ is an $\mathcal{E}$-admissible partial order, we get $p \leq q$. Since I is a down set, this in turn implies the inclusion $p \in I$. Thus, $x \in E_p \subset |I|$, and therefore $|I|$ is closed. This completes the proof of (i).

To prove (ii), we now assume that I is convex. It follows from Proposition 3.1.2 that both $I^{\leq}$ and $I^{<}$ are down sets, and due to (i), the sets $|I^{\leq}|$ and $|I^{<}|$ are closed. Moreover, since $I = I^{\leq} \setminus I^{<}$ and $\mathcal{E}$ is a partition, we have $|I| = |I^{\leq}| \setminus |I^{<}|$. Hence, the set $|I|$ is locally closed by Proposition 3.2.1. □

With any acyclic partition of a Lefschetz complex, one can associate a connection matrix. In order to make use of the theory developed in the previous chapter, we first need to recognize the Lefschetz complex as a poset filtered chain complex. Thus, assume that $\mathcal{E} = (E_p)_{p\in P}$ is an acyclic partition of a Lefschetz complex X. Then the module $C(X)$ spanned by the cells of X admits the P-gradation

$$C(X) = \bigoplus_{p\in P} C(E_p), \tag{6.2}$$

where $C(E_p) = 0$ if $E_p = \varnothing$. We have the following proposition.

Proposition 6.1.3 *The triple $(P, C(X), \partial^\kappa)$, with an $\mathcal{E}$-admissible partial order $\leq$ on P, and the P-gradation* (6.2)*, is a poset filtered chain complex.*

Proof We need to prove that ∂^κ is a filtered module homomorphism, that is, the boundary operator ∂^κ is an α-filtered module homomorphism with $\alpha = \operatorname{id}_P$. For this, we will verify (4.6) of Corollary 4.1.4 for $M = C(X)$ and $L \in \operatorname{Down}(P)$. Then one immediately obtains $M_L = C(X)_L = C(|L|)$. Hence, we need to verify that the inclusion $\partial^\kappa(C(|L|)) \subset C(|L|)$ holds. But, this follows from Proposition 3.5.7, because, by Proposition 6.1.2(i), the set $|L|$ is closed. Thus, by Corollary 4.1.4, we see that the boundary homomorphism is id_P-filtered. Therefore, $(P, C(X), \partial^\kappa)$ is a poset filtered chain complex. □

Proposition 6.1.3 enables us to define the Conley complex and connection matrix of an acyclic partition of a Lefschetz complex. For this, we consider the poset P as an object of DPSET with distinguished subset $P_\star$ satisfying (4.16), that is, we assume that $C(E_p)$ is homotopically inessential for $p \in P \setminus P_\star$.

Definition 6.1.4 By a *filtration* of an acyclic partition $\mathcal{E} = (E_p)_{p\in P}$ of a Lefschetz complex X, we mean a poset filtered chain complex $(P, C(X), \partial^\kappa)$ with P ordered by an $\mathcal{E}$-admissible partial order. The *Conley complex* and associated *connection matrix* of $\mathcal{E}$ is defined as the Conley complex and associated connection matrix of the filtration of $\mathcal{E}$, i.e., it is defined as the Conley complex and respective connection matrix of the poset filtered chain complex $(P, C(X), \partial^\kappa)$.

In some cases, it is convenient to consider an acyclic partition $\mathcal{E}$ as a self-indexed partition. Then the partial order induced by (6.1) is imposed directly on $\mathcal{E}$. Hence, the Conley complex and the associated connection matrix of a self-indexed acyclic partition $\mathcal{E}$ is the Conley complex and the respective connection matrix of $(\mathcal{E}, C(X), \partial^\kappa)$.

We would like to point out that since in this chapter we consider a general Lefschetz complex X, the existence of a Conley complex and associated connection matrix in the above definition is not guaranteed, unless we assume field coefficients in the Lefschetz complex. Nevertheless, we will see in the next section that for a more specific acyclic partition, the existence question can be settled in general.

6.2 The Singleton Partition

While a given Lefschetz complex can have many different acyclic partitions, one of the most important ones from a technical point of view is its finest one, which is given by the partition into singletons. For this special case, we have the following proposition and definition.

Proposition and Definition 6.2.1 *Assume that (X, κ) is a Lefschetz complex. Then the following hold:*

(i) The family $\mathcal{X} := \{\,\{x\} \mid x \in X\,\}$ of all singletons is a partition of X into locally closed sets.
(ii) The map sing $: X \ni x \mapsto \{x\} \in \mathcal{X}$ *is a bijection which preserves in both directions the face relation $\leq_\kappa$ on X and the relation $\preceq_{\mathcal{X}}$ on $\mathcal{X}$ given by* (6.1).
(iii) The relation $\preceq_{\mathcal{X}}$ is a partial order on $\mathcal{X}$ which coincides with the inherent partial order $\leq_{\mathcal{X}}$ on $\mathcal{X}$.
(iv) The family $\mathcal{X}$ is an acyclic partition of X.

We call $\mathcal{X}$ the singleton partition *of X.*

Proof Clearly, $\mathcal{X}$ is a partition of X and every singleton $\{x\}$ is locally closed, because we have $\{x\} = x^{\leq_\kappa} \setminus x^{<_\kappa}$ and the sets $x^{\leq_\kappa}$ and $x^{<_\kappa}$ are both down sets, hence closed. This proves (i). Now let $x, y \in X$. Then one can easily verify the sequence of equivalences

$$x \leq_\kappa y \quad \Leftrightarrow \quad x \in \operatorname{cl} y \quad \Leftrightarrow \quad \{x\} \cap \operatorname{cl}\{y\} \neq \varnothing \quad \Leftrightarrow \quad \{x\} \preceq_{\mathcal{X}} \{y\},$$

which proves (ii). Hence, $\preceq_{\mathcal{X}}$ is a partial order on $\mathcal{X}$. In particular, $\preceq_{\mathcal{X}}$ is transitively closed, and this implies that $\preceq_{\mathcal{X}}$ coincides with the inherent partial order on $\mathcal{X}$. This proves (iii). It follows that $\mathcal{X}$ is an acyclic partition of X, which establishes (iv). □

Proposition 6.2.1 shows that we have a well-defined poset filtered chain complex $(\mathcal{X}, C(X), \partial^\kappa)$. Using the order isomorphism $\operatorname{sing} : X \to \mathcal{X}$, we can then identify it with $(X, C(X), \partial^\kappa)$. Notice that via this identification, closed sets in X are in one-to-one correspondence with down sets in $\mathcal{X}$.

Proposition 6.2.2 *Let (X, κ) be a Lefschetz complex. Then the native partial order of the boundary map ∂^κ is precisely the face relation $\leq_\kappa$.*

Proof According to Definition 4.3.1, we have to prove that for every admissible partial order $\leq$ on X and arbitrary $x, y \in X$, the inequality $x \leq_\kappa y$ implies $x \leq y$. Clearly, it suffices to prove that $x \prec_\kappa y$ implies $x \leq y$, because $\leq_\kappa$ is the transitive closure of $\prec_\kappa$. Yet, the inequality $x \prec_\kappa y$ by definition implies $\kappa(y, x) \neq 0$, which in turn yields $\partial^\kappa_{yx} \neq 0$, and the admissibility of $\leq$ finally gives $x \leq y$. □

We would like to point out that the bijection sing allows us to identify partial orders in X with partial orders in the associated singleton partition $\mathcal{X}$. As an immediate consequence of Propositions 6.2.1 and 6.2.2, we get the following corollary.

Corollary 6.2.3 *Let (X, κ) be a Lefschetz complex and let $\mathcal{X}$ denote the associated singleton partition. A partial order is ∂^κ-admissible if and only if it is $\mathcal{X}$-admissible.* □

We say that a ∂^κ-admissible partial order $\leq$ in X is *natural* if

$$x \leq y \quad \text{and} \quad \dim x = \dim y \quad \Rightarrow \quad x = y. \tag{6.3}$$

Note that the native partial order of ∂^κ is natural, but that a ∂^κ-admissible partial order $\leq$ in X does not need to be natural in general.

Definition 6.2.4 (Filtration of a Lefschetz Complex) A *filtration of a Lefschetz complex* (X, κ) is defined as a filtration of the singleton partition of X, i.e., as the poset filtered chain complex $(X, C(X), \partial^\kappa)$ with X ordered by a ∂^κ-admissible partial order.

We note that among the admissible partial orders there is the native partial order (see Definition 4.3.1). It is straightforward to observe that the native partial order of $(X, C(X), \partial^\kappa)$ coincides with the face relation $\leq_\kappa$ in X. When we consider X as a poset ordered by a natural partial order in X, then we refer to the filtration $(X, C(X), \partial^\kappa)$ as a *natural filtration of* X. Finally, when we consider X as a poset ordered by the native partial order of X, then we refer to the filtration $(X, C(X), \partial^\kappa)$ as the *native filtration of* X.

Theorem 6.2.5 *Let (X, κ) be a Lefschetz complex. Then the following hold:*

(i) Every filtration of the Lefschetz complex (X, κ) is reduced, that is, it is a Conley complex of itself. In particular, $X_\star = X$.
(ii) Every natural filtration of a Lefschetz complex (X, κ) has a uniquely determined Conley complex and connection matrix. The connection matrix coincides with the (X, X)-matrix of the boundary homomorphism ∂^κ.

Proof Consider an arbitrary $x \in X$. We have from (3.7) that $\kappa_{|\{x\}\times\{x\}} = 0$. Therefore, $\partial^\kappa_{|C(\{x\})} = 0$. Moreover, since $C_{\dim x}(\{x\}) = Rx \neq 0$, we see that $X_\star = X$, which means that $(X, C(X), \partial^\kappa)$ is reduced. In consequence, it is a Conley complex of itself. This proves (i).

In order to establish assertion (ii), it suffices to verify that every transfer morphism from $(X, C(X), \partial^\kappa)$ to a Conley complex (P, C, d) of $(X, C(X), \partial^\kappa)$ is essentially graded. Actually, we will prove the stronger fact that every such transfer morphism is graded. Thus, assume that $(X, C(X), \partial^\kappa)$ is a natural filtration of X, that (P, C, d) is another Conley complex of X, and that the map $(\alpha, \varphi) : (X, C(X), \partial^\kappa) \to (P, C, d)$ is a filtered chain isomorphism.

Since in view of statement (i) the filtration $(X, C(X), \partial^\kappa)$ is a Conley complex of itself, the transfer morphism from $(X, C(X), \partial^\kappa)$ to (P, C, d) is just (α, φ). Moreover, since both $(X, C(X), \partial^\kappa)$ and (P, C, d), as Conley complexes, are reduced, we see from Proposition 5.5.4 that (α, φ) is also an isomorphism in FMOD. Therefore, it follows immediately from Lemma 4.2.5 that $\alpha : P \to X$ is an order isomorphism, and that $\varphi_{p\alpha(p)} : C(\{\alpha(p)\}) \to C_p$ is an isomorphism of $\mathbb{Z}$-graded moduli for every $p \in P$. But, Proposition 3.5.9 provides the structure of $C(\{\alpha(p)\})$. Hence, we can choose a nonzero $c_p \in C$ such that

$$(C_p)_n = \begin{cases} Rc_p & \text{if } n = \dim \alpha(p), \\ 0 & \text{otherwise.} \end{cases}$$

In other words, C_p is a one-dimensional $\mathbb{Z}$-graded module generated by c_p such that

$$\dim c_p = \dim \alpha(p). \tag{6.4}$$

In order to prove that the morphism (α, φ) is an isomorphism in GMOD, we will first show that φ is α-graded by verifying (4.1). Assume therefore that we have $\varphi_{py} \neq 0$ for $p \in P$ and $y \in X$. Furthermore, let $x = \alpha(p)$. Since (α, φ) is a filtered module homomorphism, we get from (4.2) that $p \in \alpha^{-1}(y^{\leq})^{\leq}$. This implies that there exists an element $p' \in P$ such that $p \leq p'$ and $\alpha(p') \leq y$. Since α is an order isomorphism, one further obtains $x = \alpha(p) \leq \alpha(p') \leq y$.

Note that C_p is a one-dimensional $\mathbb{Z}$-graded module which is nonzero only in dimension $n = \dim \alpha(p) = \dim x$. Since $\varphi_{py} : C(\{y\}) \to C_p$ is a $\mathbb{Z}$-graded homomorphism of degree zero, both $C(\{y\})$ and C_p are one-dimensional $\mathbb{Z}$-graded modules, and we have $\varphi_{py} \neq 0$, one immediately obtains that φ_{py} maps Ry isomorphically onto Rc_p. It follows that $\dim c_p = \dim y$. In view of $x = \alpha(p)$

and (6.4), we therefore have $\dim x = \dim y$. Since $x \leq y$ and the partial order is natural, this implies $x = y$, as well as $\alpha(p) = x = y$. This in turn proves that φ is α-graded, that is, (α, φ) is a module homomorphism in GMOD. Since it is an isomorphism in FMOD, one obtains from Corollary 4.2.6 that it is also an isomorphism in GMOD. Since it is a chain map, Proposition 3.4.1 shows that it is an isomorphism in PGCC. This completes the proof of (ii). □

We would like to stress the fact that the above result does establish the existence of the Conley complex and associated connection matrix for the singleton partition directly, i.e., without any reference to Theorem 5.3.2. In other words, in this specific situation, one does not have to assume that the Lefschetz complex X has field coefficients.

6.3 Refinements of Acyclic Partitions

We conclude this chapter with a discussion of refinements. For this, consider two objects of DPSET, that is, posets P and Q with distinguished subsets $P_\star$ and $Q_\star$, an order preserving surjection $\mu : Q \to P$ such that

$$\mu(Q_\star) = P_\star, \tag{6.5}$$

and a module M which is both P-filtered and Q-filtered, that is, we have objects (P, M) and (Q, M) in FMOD.

Definition 6.3.1 (Refinement) In the above situation, we say that (Q, M) is a μ-*refinement* of (P, M) if for every element $p \in P$ one has

$$M_p = \bigoplus_{q \in \mu^{-1}(p)} M_q. \tag{6.6}$$

Given a chain complex (C, d) which is both P-filtered and Q-filtered, we say that the poset filtered chain complex (Q, C, d) is a μ-*refinement* of the poset filtered chain complex (P, C, d) if (Q, C) is a μ-refinement of (P, C) as a module.

Proposition 6.3.2 *Assume that the object* (Q, M) *of* FMOD *is a* μ*-refinement of the object* (P, M)*, and that* $(\mathrm{id}_Q, h) : (Q, M) \to (Q, M)$ *is a morphism in* FMOD. *Then also* $(\mathrm{id}_P, h) : (P, M) \to (P, M)$ *is a morphism in* FMOD.

Proof We need to verify that $h_{p'p} \neq 0$ for $p, p' \in P$ implies $p' \leq p$. Hence, assume that $h_{p'p} \neq 0$ holds for some $p, p' \in P$. Then there exist elements $q \in \mu^{-1}(p)$ and $q' \in \mu^{-1}(p')$ such that $h_{q'q} \neq 0$. Since (id_Q, h) is a morphism in FMOD, the homomorphism h is id_Q-filtered. Therefore, we get $q' \leq q$, and since μ is order preserving, we obtain $p' = \mu(q') \leq \mu(q) = p$. This proves that also (id_P, h) is a morphism in FMOD. □

Proposition 6.3.3 *Assume that the poset filtered chain complex* (Q, C, d) *is a* μ-*refinement of the poset filtered chain complex* (P, C, d). *Then we have*

$$C_A = C_{\mu^{-1}(A)} \quad \textit{for every} \quad A \subset P. \tag{6.7}$$

Proof From the definition (3.3) of C_A and (6.6), one immediately obtains

$$C_A = \bigoplus_{p \in A} C_p = \bigoplus_{p \in A} \left(\bigoplus_{q \in \mu^{-1}(p)} C_q \right) = \bigoplus_{q \in \mu^{-1}(A)} C_q = C_{\mu^{-1}(A)}.$$

This completes the proof of the proposition. □

Consider now a poset filtered chain complex (P, C, d), as well as a Conley complex $(Q_\star, \bar{C}, \bar{d})$ of a μ-refinement (Q, C, d) of (P, C, d). In addition, define the map $\bar{\mu} := \mu \circ \iota_Q : Q_\star \to P_\star$, where $\iota_Q : Q_\star \hookrightarrow Q$ stands for the inclusion map. Notice that, in view of (6.5), the map $\bar{\mu}$ is a surjection. For $p \in P_\star$, define

$$\bar{C}_p := \bigoplus_{q \in \bar{\mu}^{-1}(p)} \bar{C}_q.$$

Since $\bar{\mu}$ is a surjection, we clearly have

$$\bar{C} = \bigoplus_{p \in P_\star} \bar{C}_p, \tag{6.8}$$

which makes $\bar{C}$ a $P_\star$-graded module.

Proposition 6.3.4 *In the above situation, the triple* $(P_\star, \bar{C}, \bar{d})$ *is a poset filtered chain complex.*

Proof Clearly, $(Q_\star, \bar{C})$ is a $\bar{\mu}$-refinement of $(P_\star, \bar{C})$. Since $(Q_\star, \bar{C}, \bar{d})$ is a poset filtered chain complex, the boundary map $\bar{d}$ is $\mathrm{id}_{Q_\star}$-filtered. By Proposition 6.3.2 it is also $\mathrm{id}_{P_\star}$-filtered. Hence, $(P_\star, \bar{C}, \bar{d})$ is indeed a poset filtered chain complex. □

The following proposition shows that, under suitable assumptions, the grouping (6.8) of components in the Conley complex $(Q_\star, \bar{C}, \bar{d})$ of (Q, C, d) provides a Conley complex of (P, C, d).

Proposition 6.3.5 *In the situation described above, assume further that in the poset filtered chain complex* $(P_\star, \bar{C}, \bar{d})$ *all of the homomorphisms* $\bar{d}_{pp}$ *are boundaryless for* $p \in P_\star$. *Then* $(P_\star, \bar{C}, \bar{d})$ *is a Conley complex of* (P, C, d).

Proof Clearly, the complex $(P_\star, \bar{C}, \bar{d})$ is peeled, and therefore, it is also reduced, because we assume that $\bar{d}_{pp}$ is boundaryless for $p \in P_\star$. Thus, we only need to verify that the poset filtered chain complex (P, C, d) is elementarily filtered equivalent to $(P_\star, \bar{C}, \bar{d})$. Since $(Q_\star, \bar{C}, \bar{d})$ is a Conley complex of (Q, C, d), there are mutually inverse elementary chain equivalences $(\iota_Q, \varphi) : (Q, C, d) \to (Q_\star, \bar{C}, \bar{d})$ and $(\iota_Q^{-1}, \psi) : (Q_\star, \bar{C}, \bar{d}) \to (Q, C, d)$.

Consider $(\iota_P, \varphi) : (P, C, d) \to (P_\star, \bar{C}, \bar{d})$ and $(\iota_P^{-1}, \psi) : (P_\star, \bar{C}, \bar{d}) \to (P, C, d)$, with $\iota_P : P_\star \hookrightarrow P$ denoting the inclusion. We claim that they are also mutually inverse elementary chain equivalences. We will first prove that (ι_P, φ) is a morphism in PFCc. It suffices to prove that it is ι_P-filtered, i.e., that (ι_P, φ) is a morphism in FMod, because, since (ι_Q, φ) is a morphism in PFCc, the homomorphism φ is already a chain map. For this, let $L \in \mathrm{Down}(P)$ be a down set. By Proposition 4.1.3, one merely needs to verify the inclusion

$$\varphi(C_L) \subset \bar{C}_{\iota_P^{-1}(L)^{\leq}}. \tag{6.9}$$

From Propositions 6.3.3 and 4.1.3 applied to φ as an ι_Q-filtered homomorphism, one obtains

$$\varphi(C_L) = \varphi(C_{\mu^{-1}(L)}) \subset \bar{C}_{\iota_Q^{-1}(\mu^{-1}(L))^{\leq}} = \bar{C}_{\bar{\mu}^{-1}(L)^{\leq}}. \tag{6.10}$$

Note that $\bar{\mu}^{-1}(L) = \bar{\mu}^{-1}(L \cap P_\star)$, since $\operatorname{dom} \bar{\mu} = Q_\star$, and, by (6.5), we have the equality $\mu(Q_\star) = P_\star$. Therefore, one further obtains

$$\bar{C}_{\bar{\mu}^{-1}(L)^{\leq}} = \bar{C}_{\bar{\mu}^{-1}(L \cap P_\star)^{\leq}}. \tag{6.11}$$

Observe now that $\bar{\mu} : Q_\star \to P_\star$ is a surjection, and $(Q_\star, \bar{C}, \bar{d})$ is a $\bar{\mu}$-refinement of $(P_\star, \bar{C}, \bar{d})$ by the very definition (6.8) of $(P_\star, \bar{C}, \bar{d})$. Hence, Proposition 6.3.3 applied to $\bar{\mu}$ yields

$$\bar{C}_{\bar{\mu}^{-1}(L \cap P_\star)^{\leq}} = \bar{C}_{(L \cap P_\star)^{\leq}} = \bar{C}_{\iota_P^{-1}(L)^{\leq}}. \tag{6.12}$$

Combining (6.10), (6.11), and (6.12), we finally get (6.9). This completes the proof that (ι_P, φ) is a morphism in PFCc.

Next we will prove that (ι_P^{-1}, ψ) is a morphism in PFCc. Again, it suffices to verify that ψ is ι_P^{-1}-filtered. Hence, assume that $L \in \mathrm{Down}(P_\star)$ is a down set. We first observe that

$$\mu^{-1}(L)^{\leq} \subset \mu^{-1}(L^{\leq}). \tag{6.13}$$

Indeed, if we assume that $q \in \mu^{-1}(L)^{\leq}$, then $q \leq q'$ for some $q' \in \mu^{-1}(L)$, which further gives $\mu(q) \leq \mu(q') \in L$ and proves that $q \in \mu^{-1}(L^{\leq})$. Hence, applying Proposition 6.3.3 and using the fact that ψ is ι_Q^{-1}-filtered, in combination with the observation that $(\iota_Q^{-1})^{-1}(A) = A$ for $A \subset Q_\star$, we get

$$\psi(\bar{C}_L) = \psi(\bar{C}_{\bar{\mu}^{-1}(L)}) \subset C_{(\iota_Q^{-1})^{-1}(\bar{\mu}^{-1}(L))^{\leq}} = C_{\bar{\mu}^{-1}(L)^{\leq}} = C_{\iota_Q^{-1}(\mu^{-1}(L))^{\leq}}.$$

Clearly, one has the inclusion $\iota_Q^{-1}(\mu^{-1}(L)) = Q_\star \cap \mu^{-1}(L) \subset \mu^{-1}(L)$. In combination with (6.13), and again Proposition 6.3.3, this further shows that

$$C_{\iota_Q^{-1}(\mu^{-1}(L))^{\leq}} \subset C_{\mu^{-1}(L)^{\leq}} \subset C_{\mu^{-1}(L^{\leq})} = C_{L^{\leq}} = C_{(\iota_P^{-1})^{-1}(L^{\leq})}.$$

This proves that ψ is ι_P^{-1}-filtered and (ι_P^{-1}, ψ) is a morphism in PFCC.

Finally, consider an elementary filtered chain homotopy T between

$$(\iota_Q^{-1}, \psi) \circ (\iota_Q, \varphi) = (\mathrm{id}_{Q_\star}, \psi\varphi) : (Q, C, d) \to (Q, C, d) \quad \text{and}$$
$$\mathrm{id}_{(Q,C,d)} = (\mathrm{id}_Q, \mathrm{id}_C) : (Q, C, d) \to (Q, C, d).$$

By Proposition 6.3.2, the homomorphism T is id_P-filtered. Therefore, it is straightforward to see that T is also an elementary filtered chain homotopy between

$$(\iota_P^{-1}, \psi) \circ (\iota_P, \varphi) = (\mathrm{id}_{P_\star}, \psi\varphi) : (P, C, d) \to (P, C, d) \quad \text{and}$$
$$\mathrm{id}_{(P,C,d)} = (\mathrm{id}_P, \mathrm{id}_C) : (P, C, d) \to (P, C, d).$$

Similarly, we find an elementary chain homotopy between

$$(\iota_P, \varphi) \circ (\iota_P^{-1}, \psi) = (\mathrm{id}_{P_\star}, \varphi\psi) : (P_\star, \bar{C}, \bar{d}) \to (P_\star, \bar{C}, \bar{d}) \quad \text{and}$$
$$\mathrm{id}_{(P_\star,\bar{C},\bar{d})} = (\mathrm{id}_{P_\star}, \mathrm{id}_{\bar{C}}) : (P_\star, \bar{C}, \bar{d}) \to (P_\star, \bar{C}, \bar{d}).$$

This proves that $(P_\star, \bar{C}, \bar{d})$ is a Conley complex of (P, C, d). □

Proposition 6.3.5 facilitates the computation of Conley complexes of acyclic partitions of Lefschetz complexes.

Definition 6.3.6 (Refinements of Acyclic Partitions) Let $\mathcal{F}$ and $\mathcal{E}$ be two self-indexed acyclic partitions of a Lefschetz complex X. Then $\mathcal{F}$ is a *refinement* of $\mathcal{E}$ if for every $F \in \mathcal{F}$ there exists an $E \in \mathcal{E}$ such that $F \subset E$.

Assume that the acyclic partition $\mathcal{F}$ of X is a refinement of the acyclic partition $\mathcal{E}$ of X. Then, for every $F \in \mathcal{F}$, there is exactly one $E \in \mathcal{E}$ such that $F \subset E$. Therefore, we have a well-defined map

$$\mu = \mu_{\mathcal{F},\mathcal{E}} : \mathcal{F} \ni F \mapsto E \in \mathcal{E}.$$

Since $\mathcal{E}$ and $\mathcal{F}$ are partitions, the map μ is clearly a surjection. It is straightforward to verify that μ preserves the inherent partial orders of $\mathcal{F}$ and $\mathcal{E}$. Therefore, we have the following proposition.

Proposition 6.3.7 *Consider two acyclic partitions $\mathcal{F}$ and $\mathcal{E}$ of a Lefschetz complex X as objects of* DPSET, *with the distinguished subsets $\mathcal{F}_\star$ and $\mathcal{E}_\star$ satisfying* (4.16). *If $\mathcal{F}$ is a refinement of $\mathcal{E}$ and $\mu_{\mathcal{F},\mathcal{E}}(\mathcal{F}_\star) = \mathcal{E}_\star$, then the filtration $(\mathcal{F}, C(X), \partial^\kappa)$ of X is a $\mu_{\mathcal{F},\mathcal{E}}$-refinement of the filtration $(\mathcal{E}, C(X), \partial^\kappa)$.* □

Chapter 7
Dynamics of Combinatorial Multivector Fields

Multivector fields are a natural generalization of Forman's combinatorial vector fields. They provide greater flexibility in the description of a variety of dynamical phenomena, and they can be used to combinatorialize even such concepts as chaotic behavior or multiflows. In the present chapter, we review basic notions from combinatorial multivector fields on Lefschetz complexes and show that they provide a natural framework for our theory of connection matrices. This is achieved through a straightforward construction of an associated acyclic partition of the underlying Lefschetz complex.

7.1 Combinatorial Multivector Fields

The concept of a combinatorial multivector field on a general Lefschetz complex was originally proposed in [35, Definition 5.10] as a dynamically oriented extension of the notion of combinatorial vector field in the sense of Forman [18, 19]. The definition of combinatorial multivector field considered here is based on [30], yet again restricted to the special setting of Lefschetz complexes. Thus, let X be an arbitrary Lefschetz complex. By a *combinatorial multivector* in X, we mean any non-empty subset of X which is locally closed with respect to the Lefschetz topology. Then a *combinatorial multivector field* is a self-indexed partition $\mathcal{V}$ of X into combinatorial multivectors. Since in this book we do not use any other concepts of multivector or vector fields, in the sequel, we simplify the terminology by dropping the adjective combinatorial in combinatorial multivector and combinatorial multivector field.

There are two distinct types of multivectors. We say that a multivector V is *critical*, if the relative Lefschetz homology $H(\operatorname{cl} V, \operatorname{mo} V)$ is nonzero. A multivector V which is not critical is called *regular*. Recall that in both cases, the relative homology equals the Lefschetz homology $H(V)$, see Proposition 3.5.8. For each

M. Mrozek, T. Wanner, *Connection Matrices in Combinatorial Topological Dynamics*, SpringerBriefs in Mathematics,
https://doi.org/10.1007/978-3-031-87600-4_7

element $x \in X$, we denote by $[x]_{\mathcal{V}}$ the unique multivector in $\mathcal{V}$ which contains x. If the multivector field $\mathcal{V}$ is clear from context, then we abbreviate the notation by writing $[x] := [x]_{\mathcal{V}}$. In addition, we say that $x \in X$ is *critical* (respectively, *regular*) with respect to the multivector field $\mathcal{V}$, if the multivector $[x]_{\mathcal{V}}$ is critical (respectively, regular). Finally, a subset $A \subset X$ is called *$\mathcal{V}$-compatible* if it is the union of multivectors. More precisely, the set A is $\mathcal{V}$-compatible if for every element $x \in X$ one either has the inclusion $[x]_{\mathcal{V}} \subset A$ or the equality $[x]_{\mathcal{V}} \cap A = \varnothing$.

We associate with every multivector field a multivalued map $\Pi_{\mathcal{V}} : X \multimap X$ given by the definition

$$\Pi_{\mathcal{V}}(x) := \operatorname{cl} x \cup [x]_{\mathcal{V}}. \tag{7.1}$$

The multivalued map $\Pi_{\mathcal{V}}$ may be interpreted as a digraph with vertices in X and an arrow from $x \in X$ to $y \in X$ if $y \in \Pi_{\mathcal{V}}(x)$. Clearly, every multivector $V \in \mathcal{V}$ forms a clique in this digraph. By collapsing all vertices in a multivector to a point, we obtain an induced digraph with vertices in $\mathcal{V}$ and an arrow from $V \in \mathcal{V}$ to $W \in \mathcal{W}$ if there exist an $x \in V$ and a $y \in W$ such that $y \in \Pi_{\mathcal{V}}(x)$. We refer to this digraph as the *$\mathcal{V}$-digraph.*

Since $\mathcal{V}$ is a self-indexed partition of X into locally closed subsets, we can consider the relation $\preceq_{\mathcal{V}}$ in $\mathcal{V}$ introduced in (6.1) at the beginning of Sect. 6.1. The following proposition shows that the $\mathcal{V}$-digraph and the relation $\preceq_{\mathcal{V}}$ are the same concepts.

Proposition 7.1.1 *There is an arrow from V to W in the $\mathcal{V}$-digraph of $\mathcal{V}$ if and only if $W \preceq_{\mathcal{V}} V$, that is, if and only if we have $W \cap \operatorname{cl} V \neq \varnothing$.*

Proof We begin by assuming that there is an arrow from V to W in the $\mathcal{V}$-digraph of the multivector field $\mathcal{V}$, and we choose elements $x, y \in X$ in such a way that both $x \in V$ and $y \in W$, as well as $y \in \Pi_{\mathcal{V}}(x) = [x]_{\mathcal{V}} \cup \operatorname{cl} x$, are satisfied. If one further has $y \in [x]$, then $W = [y] = [x] = V$, and this yields $W \cap \operatorname{cl} V = V \neq \varnothing$. If on the other hand $y \in \operatorname{cl}[x]$, then $y \in W \cap \operatorname{cl} V$, and therefore $W \cap \operatorname{cl} V \neq \varnothing$.

To establish the opposite implication, suppose that $W \cap \operatorname{cl} V \neq \varnothing$. Then we can take a $y \in W \cap \operatorname{cl} V$ and an $x \in V$ such that $y \in \operatorname{cl} x$. It follows that $y \in \Pi_{\mathcal{V}}(x)$. Hence, there is an arrow in the $\mathcal{V}$-digraph from V to W. □

Example 7.1.2 (Nonuniqueness via Subdivision, ◁ 5.5.7 ▷ 7.3.2) Figure 7.1 presents three different combinatorial multivector fields $\mathcal{V}_0$, $\mathcal{V}_1$, and $\mathcal{V}_2$ on the Lefschetz complex X introduced in Example 3.5.6. The $\mathcal{V}_0$-digraph coincides with the Hasse diagram (4.14). Notice that by using the Hasse diagram representation of the digraph, we implicitly assume that arrows always point downward, i.e., we represent them without arrow heads. Moreover, to keep the diagrams as simple as

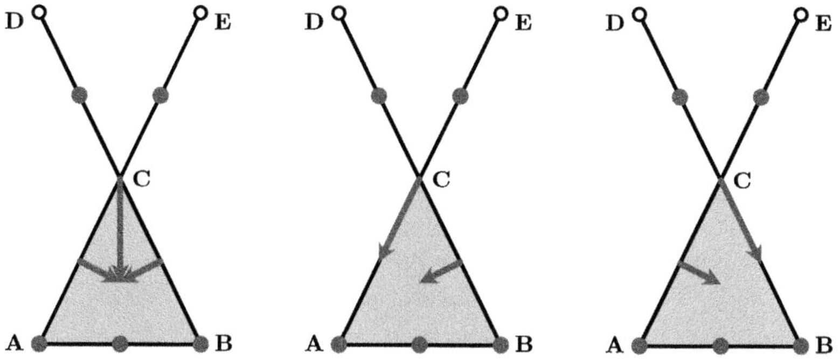

Fig. 7.1 Three different multivector fields $\mathcal{V}_0$ (left), $\mathcal{V}_1$ (middle), and $\mathcal{V}_2$ (right) on the Lefschetz complex X introduced in Example 3.5.6. Critical cells are marked with a fat circle in the center of mass of a simplex. Vectors and multivectors are indicated by an arrow from simplex x to simplex y whenever $y \in \Pi_\mathcal{V}(x)$ and $y \notin \mathrm{cl}\, x$, and as long as y is a top-dimensional coface in the multivector containing x. The remaining cases of arrows $x \to y$ are not marked in order to keep the image readable

possible, we do not indicate the loops which are present at every node. Similarly, the $\mathcal{V}_1$-digraph is

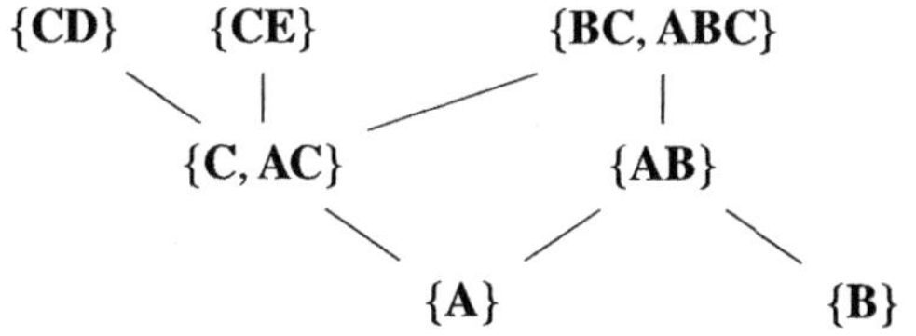

$$\begin{array}{ccc}
\{CD\} \quad \{CE\} & & \{BC, ABC\} \\
\searrow \quad | \quad \swarrow & & | \\
\{C, AC\} & & \{AB\} \\
\searrow & \swarrow \quad \searrow & \\
\{A\} & & \{B\}
\end{array} \qquad (7.2)$$

and the $\mathcal{V}_2$-digraph is given by

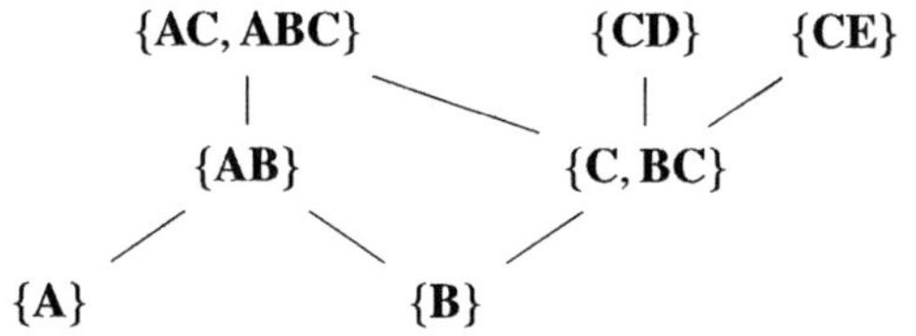

$$\begin{array}{ccc}
\{AC, ABC\} & \{CD\} \quad \{CE\} \\
| & \searrow \quad | \quad \swarrow \\
\{AB\} & \{C, BC\} \\
\swarrow \quad \searrow & \swarrow \\
\{A\} \quad \{B\} &
\end{array} \qquad (7.3)$$

We will return to these three multivector fields later in more detail. ◊

We call a subset $A \subset \mathbb{Z}$ of the integers *left bounded* (respectively, *right bounded*) if it has a minimum (respectively, maximum). Otherwise, we call it *left unbounded* or *left infinite* (respectively, *right unbounded* or *right infinite*). We call $A \subset \mathbb{Z}$ *bounded* if A has both a minimum and a maximum. Finally, we call $A \subset \mathbb{Z}$ a $\mathbb{Z}$-*interval* if we have $A = \mathbb{Z} \cap I$ where I is an interval in $\mathbb{R}$.

A *solution* of a multivector field $\mathcal{V}$ in $A \subset X$ is a partial map $\varrho : \mathbb{Z} \nrightarrow A$ whose *domain*, denoted by $\mathrm{dom}\,\varrho$, is a $\mathbb{Z}$-interval and such that for arbitrary $i, i+1 \in \mathrm{dom}\,\varrho$

the inclusion $\varrho(i+1) \in \Pi_{\mathcal{V}}(\varrho(i))$ is satisfied. We say that the solution *passes through* $x \in X$ if $0 \in \operatorname{dom} \varrho$ and $x = \varrho(0)$. The solution ϱ is *full* if $\operatorname{dom} \varrho = \mathbb{Z}$. It is *periodic* if there is a $k > 0$ such that $\varrho(i+k) = \varrho(i)$ for all $i \in \mathbb{Z}$. It is a *partial solution* or simply a *path* if $\operatorname{dom} \varrho$ is bounded. We refer to the cardinality of the domain of a path as the *length* of the path. If the maximum of $\operatorname{dom} \varrho$ exists, we call the value of ϱ at this maximum the *right endpoint* of ϱ. If the minimum of $\operatorname{dom} \varrho$ exists, we call the value of ϱ at this minimum the *left endpoint* of ϱ. We denote the left and right endpoints of ϱ by $\varrho^{\sqsubset}$ and $\varrho^{\sqsupset}$, respectively. Given a full solution ϱ through $x \in X$, we let $\varrho^+ := \varrho_{|\mathbb{Z}_0^+}$ denote the *forward solution* through x, and the *backward solution* through x is denoted by $\varrho^- := \varrho_{|\mathbb{Z}_0^-}$.

By a *shift* of a solution ϱ, we mean the composition $\varrho \circ \tau_n$, where the translation map is defined as $\tau_n : \mathbb{Z} \ni m \mapsto m + n \in \mathbb{Z}$. Given two solutions φ and ψ such that the left endpoint $\psi^{\sqsubset}$ and the right endpoint $\varphi^{\sqsupset}$ exist, and such that $\psi^{\sqsubset} \in \Pi_{\mathcal{V}}(\varphi^{\sqsupset})$, then there is a unique shift τ_n such that $\varphi \cup \psi \circ \tau_n$ is a solution. We call this union of paths the *concatenation* of φ and ψ and denote it $\varphi \cdot \psi$. We also identify each element $x \in X$ with the path of length one whose image is $\{x\}$. The following proposition is straightforward.

Proposition 7.1.3 *Let $x, y \in X$. Then $y \in \Pi_{\mathcal{V}}(x)$ if and only if there is an arrow from $[x]$ to $[y]$ in the $\mathcal{V}$-digraph. Consequently, the multivector field $\mathcal{V}$ admits a path from x to y if and only if there is a path from $[x]$ to $[y]$ in the $\mathcal{V}$-digraph.* □

Finally, and for later reference, we define the *backward and forward ultimate image* of a full solution $\varrho : \mathbb{Z} \to X$, respectively, as the sets

$$\operatorname{uim}^-(\varrho) := \bigcap_{t \in \mathbb{Z}^-} \varrho((-\infty, t]),$$

$$\operatorname{uim}^+(\varrho) := \bigcap_{t \in \mathbb{Z}^+} \varrho([t, \infty)).$$

One can easily see that both of these sets are necessarily nonempty. Moreover, the following proposition is straightforward.

Proposition 7.1.4 *If $\varrho : \mathbb{Z} \to X$ is a periodic solution, then its backward and forward ultimate images satisfy* $\operatorname{uim}^-(\varrho) = \operatorname{im} \varrho = \operatorname{uim}^+(\varrho)$. □

While the above notion of solution is a natural generalization of the classical case, the specific definition of the multivalued map $\Pi_{\mathcal{V}}$ causes some slight complication. Since we clearly have $x \in \Pi_{\mathcal{V}}(x)$ for every $x \in X$, every point in the Lefschetz complex lies on a constant full solution through itself. Thus, if we are interested in extending notions such as invariance from the classical situation, our concept of solution implies that every subset of X would be an invariant set, which clearly is not appropriate.

This can be remedied by strengthening our notion of solutions to the concept of *essential solutions*. A full solution $\varrho : \mathbb{Z} \to X$ is called *left essential* (or, respectively, *right essential*), if for every regular $x \in \operatorname{im} \varrho$ the set $\{t \in \mathbb{Z} \mid \varrho(t) \notin [x]_{\mathcal{V}}\}$

is left infinite (or, respectively, right infinite). We say that ϱ is an *essential solution* if it is both left and right essential. In other words, essential solutions have to leave every regular multivector in forward and backward time before they can reenter it.

7.2 Conley Index and Morse Decompositions

After the preparations of the last section, we now turn our attention to invariance and Conley theory. We say that $S \subset X$ is *$\mathcal{V}$-invariant*, or briefly *invariant*, if for every $x \in S$ there exists an essential solution through x in S. A closed set $N \subset X$ is an *isolating set* for a $\mathcal{V}$-invariant subset $S \subset N$ if $\Pi_{\mathcal{V}}(S) \subset N$ and any path in N with endpoints in S is a path in S. Finally, we say that S is an *isolated invariant set* if S admits an isolating set. One can show that isolated invariant sets have an easy direct characterization. This is the subject of the following proposition (see [30, Propositions 4.10, 4.12, and 4.13]).

Proposition 7.2.1 (Characterization of Isolated Invariant Sets) *A Lefschetz complex subset $S \subset X$ is an isolated invariant set for the multivector field $\mathcal{V}$ if and only if it is an invariant set which is both $\mathcal{V}$-compatible and locally closed.* □

Consider an isolated invariant set S of a combinatorial multivector field $\mathcal{V}$ on a Lefschetz complex X. The Conley index of S is defined in [30, Section 5.2] as the homology of any index pair of S, in particular the homology $H(\operatorname{cl} S, \operatorname{mo} S)$ of the special index pair $(\operatorname{cl} S, \operatorname{mo} S)$ (see [30, Theorem 5.3]). Since, in view of Propositions 3.5.8 and 7.2.1, the homology $H(\operatorname{cl} S, \operatorname{mo} S)$ is isomorphic to $H(S)$, for the purposes of this book, it suffices to assume that the *Conley index of an isolated invariant set* S is

$$\operatorname{Con}(S) := H(S), \tag{7.4}$$

that is, the Lefschetz homology of S.

Observe that given an isolated invariant set S for a multivector field $\mathcal{V}$ on a Lefschetz complex X, by Proposition 7.2.1, the family

$$\mathcal{V}_S := \{ V \in \mathcal{V} \mid V \subset S \}$$

is a partition of S and, in consequence, a multivector field on S considered as a Lefschetz subcomplex of X. We call it the multivector field *induced by $\mathcal{V}$ on S*.

Multivector fields on Lefschetz complexes can exhibit a variety of different dynamical behaviors, ranging from critical cells which correspond to equilibrium solutions, to both transient and more complicated recurrent behavior. In this context, an essential, and therefore, full solution $\varrho : \mathbb{Z} \to X$ is called *recurrent* if for every element $x \in \operatorname{im} \varrho$ the set $\varrho^{-1}(x)$ is both left and right infinite. Examples of recurrent solutions include constant solutions whose image is a single critical multivector, or nonconstant periodic solutions which loop through a sequence of regular multivectors.

In classical dynamics, one has the notion of a gradient vector field, which rules out the existence of nonconstant recurrent solutions—and in fact, Forman's original combinatorial vector fields were motivated by precisely this situation. While our definition of combinatorial multivector fields is considerably more general and allows for recurrence, we will see that assuming an additional gradient structure can provide more detailed information about the associated connection matrices.

Thus, we call a multivector field $\mathcal{V}$ on a Lefschetz complex X *gradient-like* if for every recurrent solution $\varrho : \mathbb{Z} \to X$ there exists a multivector $V \in \mathcal{V}$ such that the inclusion $\operatorname{im} \varrho \subset V$ is satisfied. Moreover, a multivector field $\mathcal{V}$ is called a *gradient multivector field* if the only recurrent solutions are constant solutions whose value consists of a singleton in $\mathcal{V}$. We would like to point out that this rules out multivectors of size at least two which have nontrivial Conley index. Moreover, recall that a singleton in $\mathcal{V}$ is always critical, see Proposition 3.5.9.

Gradient-like multivector fields on Lefschetz complexes are of significant importance for our theory of connection matrices, since they correspond precisely to acyclic partitions. This is the subject of the following result.

Proposition 7.2.2 (Acyclic Partitions via Gradient-Like Multivector Fields) *Let $\mathcal{V}$ be a combinatorial multivector field on a Lefschetz complex X. The following properties are equivalent:*

(i) The multivector field $\mathcal{V}$ is gradient-like.
(ii) The $\mathcal{V}$-digraph is acyclic.
(iii) The collection $\mathcal{V}$ of multivectors is an acyclic partition of X.

Proof The equivalence (i) $\Leftrightarrow$ (ii) follows from Proposition 7.1.3, and the equivalence (ii) $\Leftrightarrow$ (iii) follows from Propositions 7.1.1 and 3.1.1. □

In order to define connection matrices for multivector fields $\mathcal{V}$ on a Lefschetz complex X, we need the concept of a Morse decomposition of an isolated invariant set.

Definition 7.2.3 (Morse Decomposition) Let S denote an isolated invariant set of a multivector field $\mathcal{V}$ on a Lefschetz complex X. Then a *Morse decomposition of S* is given by an indexed collection $\mathcal{M} = (M_p)_{p \in P}$ of mutually disjoint isolated invariant subsets of S, which are referred to as *Morse sets*, together with a partial order $\leq$ on P such that for every essential solution $\varrho : \mathbb{Z} \to S$ one either has $\operatorname{im} \varrho \subset M_p$ for some index $p \in P$, or there exist $p^{\pm} \in P$ with $p^+ < p^-$ such that both $\operatorname{uim}^-(\varrho) \subset M_{p^-}$ and $\operatorname{uim}^+(\varrho) \subset M_{p^+}$ are satisfied. By a *Morse decomposition of $\mathcal{V}$*, we mean a Morse decomposition of the maximal invariant subset of $\mathcal{V}$ in X.

We do not require the Morse sets to be non-empty, since a Morse set may be given implicitly as the maximal invariant subset in a given neighborhood, and such a set can be empty. This is where the concept of an indexed family is helpful, because two otherwise equal empty Morse sets can be distinguished by different indices. If all Morse sets are non-empty, then it is sometimes convenient to consider $\mathcal{M}$ as a self-indexed family. In this case, the poset is the family itself, and the partial order is imposed directly on $\mathcal{M}$.

A natural example of a self-indexed Morse decomposition is provided by the following proposition which follows easily from Propositions 7.2.1 and 7.2.2.

Proposition 7.2.4 *The family C of critical multivectors of a gradient-like multivector field $\mathcal{V}$ partially ordered by $\leq_{\mathcal{V}}$ is a Morse decomposition of $\mathcal{V}$.* □

We would like to point out that Definition 7.2.3 of Morse decomposition differs in two aspects from the one given in [30]. The definition in [30], instead of the condition with $\operatorname{uim}^-(\varrho)$ and $\operatorname{uim}^+(\varrho)$, uses the condition with α- and ω-limit sets of ϱ. The reader familiar with [30] will, however, immediately notice that these conditions are equivalent, in view of Proposition 7.2.1. Also, the paper [30] defines a *global Morse decomposition*, i.e., a Morse decomposition of the whole space X instead of an isolated invariant set S. Since the definition in [30] is given under the assumption that X is invariant, one can easily regain the definition of Morse decomposition of an isolated invariant set S from the definition in [30] by replacing X with S and $\mathcal{V}$ with the multivector field $\mathcal{V}_S$ induced by $\mathcal{V}$ on S. Global Morse decompositions of a combinatorial multivector field can readily be determined via the strongly connected components of the associated $\mathcal{V}$-digraph. In fact, in contrast to the classical case, there always exists a finest Morse decomposition for a combinatorial multivector field $\mathcal{V}$.

As we already mentioned, isolated invariant sets may be given implicitly as maximal invariant subsets in a given neighborhood. This, in particular, happens in computational applications, because there may be areas where it is impossible to decide whether they contain a non-empty invariant set due to limited data or computational power. As a remedy to such situations, we further define connection matrices for block decompositions, a concept slightly more general than Morse decompositions.

Definition 7.2.5 (Block Decomposition) Let $\mathcal{V}$ denote a combinatorial multivector field on a Lefschetz complex X and let $S \subset X$ be an isolated invariant set. Then a *block* in S is a locally closed and $\mathcal{V}$-compatible subset of S. A *block decomposition* of S is an indexed family $\mathcal{M} = (M_p)_{p\in P}$ of mutually disjoint blocks in S, together with a partial order $\leq$ on P and such that:

(BD1) If τ is a path in S with $\tau^{\sqsubset} \in M_{p^-}$ and $\tau^{\sqsupset} \in M_{p^+}$ for some $p^{\pm} \in P$, then we have $p^- \geq p^+$, and additionally, $p^- = p^+$ implies $\operatorname{im}\tau \subset M_{p^+} = M_{p^-}$.

(BD2) For every essential solution $\varrho \in S$, the set $\varrho^{-1}\left(\bigcup \mathcal{M}\right)$ is left and right infinite.

It follows from Proposition 7.2.1 that every isolated invariant set is a block. Note, however, that a block does not have to be an isolated invariant set, because it may fail to be invariant. But, if B is a block, then one can easily verify that

$$\operatorname{Inv} B := \{ x \in B \mid \text{there is an essential solution through } x \text{ in } B \}$$

is invariant, locally closed, and $\mathcal{V}$-compatible, hence an isolated invariant set, again by Proposition 7.2.1. Nevertheless, the set $\operatorname{Inv} B$ could very well be the empty set. We leave the proof of the following simple proposition to the reader.

Proposition 7.2.6 *If $\mathcal{M}$ is a Morse decomposition of S, then it is also a block decomposition of S. Conversely, if $\mathcal{M} = (M_p)_{p \in P}$ is a block decomposition of S, then*

$$\mathcal{M}^{\bullet} := \{ \operatorname{Inv} M_p \mid p \in P \} \tag{7.5}$$

is a Morse decomposition of S. □

A block decomposition $\mathcal{M}$ shares with a Morse decomposition its crucial property, as the following proposition shows.

Proposition 7.2.7 *Assume that the indexed collection $\mathcal{M} = \{M_p\}_{p \in P}$ is a block decomposition of an isolated invariant set S. Then for every essential solution ϱ in S, there exist $p^{\pm} \in P$ such that $p^- \geq p^+$, as well as*

$$\operatorname{uim}^- \varrho \subset M_{p^-} \quad \textit{and} \quad \operatorname{uim}^+ \varrho \subset M_{p^+}. \tag{7.6}$$

Moreover, if $p^- = p^+$, then one also has

$$\operatorname{im} \varrho \subset M_{p^-}. \tag{7.7}$$

Proof It follows from (BD1) that

$$P_\varrho := \{ p \in P \mid M_p \cap \operatorname{im} \varrho \neq \varnothing \}$$

is a linearly ordered subset of P. Set $p^- := \max P_\varrho$ and $p^+ := \min P_\varrho$, and choose indices $k^{\pm} \in \mathbb{Z}$ such that $\varrho(k^-) \in M_{p^-}$ and $\varrho(k^+) \in M_{p^+}$. We claim that

$$\varrho([k^+, \infty)) \subset M_{p^+}. \tag{7.8}$$

To see this, assume to the contrary that $\varrho(n) \notin M_{p^+}$ for some $n > k^+$. Since, by (BD2), the preimage $\varrho^{-1}\left(\bigcup \mathcal{M}\right)$ is right infinite, there exists an index $m > n$ such that $\varrho(m) \in \bigcup \mathcal{M}$. Then $\varrho(m) \in M_{p^+}$, because otherwise we contradict the definition of p^+. Consider now the path $\tau := \varrho_{|[k^+, m]}$. We have $\tau^{\sqsubset} = \varrho(k^+) \in M_{p^+}$

and $\tau^{\sqsupset} = \varrho(m) \in M_{p^+}$. Hence, we get from (BD1) that $\varrho(n) = \tau(n) \in \operatorname{im} \tau \subset M_{p^+}$, a contradiction proving (7.8). Similarly, we prove that

$$\varrho((-\infty, k^-]) \subset M_{p^-}. \tag{7.9}$$

Consider now the case $k^- \geq k^+$. Applying (BD1) to $\sigma := \varrho_{|[k^+, k^-]}$, we obtain the inequality $p^+ \geq p^-$, which implies $p^+ = p^-$, since $p^+ = \min P_\varrho \leq \max P_\varrho = p^-$. We also get from (7.8) and (7.9) that

$$\operatorname{im} \varrho \subset \varrho((-\infty, k^-]) \cup \varrho([k^+, \infty)) \subset M_{p^-},$$

which shows that ϱ satisfies (7.7). Hence, the solution ϱ also satisfies (7.6), because we have both the inclusion $\operatorname{uim}^- \varrho \subset \operatorname{im} \varrho$ and $\operatorname{uim}^+ \varrho \subset \operatorname{im} \varrho$. This completes the proof in the case $k^- \geq k^+$.

Consider finally the remaining case $k^- < k^+$. Then, again by (BD1), one has the inequality $p^- \geq p^+$, together with $\operatorname{uim}^- \varrho \subset \varrho((-\infty, k^-]) \subset M_{p^-}$ by (7.9), as well as $\operatorname{uim}^+ \varrho \subset \varrho([k^+, \infty)) \subset M_{p^+}$ by (7.8). Hence, the solution ϱ satisfies (7.6). Moreover, if $p^- = p^+$, then one has

$$\operatorname{im} \varrho \subset \varrho((-\infty, k^-]) \cup \varrho([k^-, k^+]) \cup \varrho([k^+, \infty)) \subset M_{p^+},$$

which shows that ϱ satisfies (7.6) when $p^- = p^+$. □

Corollary 7.2.8 *Assume that the indexed collection $\mathcal{M} = \{M_p\}_{p \in P}$ is a block decomposition of an isolated invariant set S. Then every multivector $V \notin \bigcup \mathcal{M}$ is a regular multivector.*

Proof Assume, to the contrary, that $V \notin \bigcup \mathcal{M}$ is critical. Let ϱ denote a constant, full solution through an element $x \in V$. Then ϱ is an essential solution, and we have the equality $\operatorname{uim}^- \varrho = \operatorname{uim}^+ \varrho = \{x\} = \operatorname{im} \varrho$. Hence, by Proposition 7.2.7, there exists an index $p \in P$ such that $\operatorname{im} \varrho \subset M_p$. It follows that $V \subset M_p$, a contradiction. □

Definition 7.2.9 (Induced Partition) Consider a Morse decomposition or a block decomposition $\mathcal{M} = (M_p)_{p \in P}$ of the multivector field $\mathcal{V}$. Set

$$\mathcal{M}^\perp := \{ V \in \mathcal{V} \mid V \cap M_p = \varnothing \text{ for all } p \in P \}.$$

Choose an extension $\hat{P} \supset P$ and an indexed family $(E_p)_{p \in \hat{P}}$ such that $E_p = M_p$ for arbitrary $p \in P$ and that each $V \in \mathcal{M}^\perp$ equals E_q for a unique $q \in \hat{P} \setminus P$. Such a family is a partition of X uniquely determined up to the choice of indexing. We call the collection $\mathcal{E} = (E_p)_{p \in \hat{P}}$ the *partition induced by the decomposition $\mathcal{M}$*. In case of potential confusion, we write $\mathcal{E}_{\mathcal{M},\mathcal{V}}$ or $\mathcal{E}_\mathcal{M}$ for $\mathcal{E}$, and $\hat{P}_{\mathcal{M},\mathcal{V}}$ or $\hat{P}_\mathcal{M}$ for $\hat{P}$.

For $p, p' \in \hat{P}$, we write $p \preccurlyeq p'$ if $p = p'$ or there is a path σ such that $\sigma^{\sqsubset} \in E_{p'}$ and $\sigma^{\sqsupset} \in E_p$. Then the following statement holds.

Proposition 7.2.10 *The above-defined relation $\preccurlyeq$ is a partial order in $\hat{P}$.*

Proof Clearly, the relation is reflexive and transitive. To see that it is antisymmetric, let $p, p' \in \hat{P}$ be such that $p \preccurlyeq p'$ and $p' \preccurlyeq p$. We have to verify that $p = p'$. Let σ be a path such that $\sigma^{\sqsubset} \in E_{p'}$ and $\sigma^{\sqsupset} \in E_p$. In addition, let τ denote a path which satisfies both $\tau^{\sqsubset} \in E_p$ and $\tau^{\sqsupset} \in E_{p'}$.

Consider first the case when $E_p, E_{p'} \in \mathcal{M}^{\perp}$. Then we have $[\sigma^{\sqsubset}] = E_{p'} = [\tau^{\sqsupset}]$ and $[\sigma^{\sqsupset}] = E_p = [\tau^{\sqsubset}]$. It follows that the paths σ and τ can be concatenated to a full, periodic solution ϱ. Assume now that we have the inequality $p \neq p'$. Then the solution ϱ is essential. Hence, by Proposition 7.2.7, we can find $p^{\pm} \in P$ such that $p^- \geq p^+$, as well as $\operatorname{uim}^- \varrho \subset M_{p^-}$, and $\operatorname{uim}^+ \varrho \subset M_{p^+}$. However, it follows from Proposition 7.1.4 that $\operatorname{uim}^-(\varrho) = \operatorname{im} \varrho = \operatorname{uim}^+(\varrho)$. Therefore, we must have the equality $p^- = p^+$. In consequence, $\sigma^{\sqsubset} \in E_{p'} \cap M_{p^-}$, which contradicts our assumption that $E_p, E_{p'} \in \mathcal{M}^{\perp}$. Thus, $p = p'$ in this case.

Consider now the case when both $E_p = M_p$ and $E_{p'} = M_{p'}$ are in $\mathcal{M}$. Then we have $p, p' \in P$. Hence, applying (BD1) to σ, we obtain the inequality $p' \geq p$, and after applying (BD1) to τ, we further get $p \geq p'$. It follows that $p = p'$.

Finally, consider the case when one of the two sets is in $\mathcal{M}$ and the other is in $\mathcal{M}^{\perp}$. Without loss of generality, we may assume that $E_p = M_p \in \mathcal{M}$ and $E_{p'} \in \mathcal{M}^{\perp}$. Then we have $[\sigma^{\sqsubset}] = E_{p'} = [\tau^{\sqsupset}]$. It follows that the paths σ and τ can be concatenated to a path π such that $\pi^{\sqsubset} = \tau^{\sqsubset} \in M_p$ and $\pi^{\sqsupset} = \sigma^{\sqsupset} \in M_p$. Applying (BD1) to π, we obtain the inclusion $\operatorname{im} \pi \subset M_p = E_p$. Since $\sigma^{\sqsubset} \in \operatorname{im} \pi$ and $\sigma^{\sqsubset} \in E_{p'}$, this readily gives $E_p \cap E_{p'} \neq \varnothing$, which implies $p = p'$, because $\mathcal{E}_{\mathcal{M}}$ is a partition. □

Since $\mathcal{E}_{\mathcal{M}}$ is a partition of the Lefschetz complex X, one can also consider the relation $\preceq_{\mathcal{E}_{\mathcal{M}}}$ defined in (6.1). As the following result demonstrates, its transitive closure is the partial order $\preccurlyeq$ from above.

Proposition 7.2.11 *The relation $\preccurlyeq$ is the transitive closure of the relation $\preceq_{\mathcal{E}_{\mathcal{M}}}$ defined in* (6.1). *Thus, the two relations $\preccurlyeq$ and $\preceq_{\mathcal{E}_{\mathcal{M}}}$ coincide, and in particular, the family $\mathcal{E}_{\mathcal{M}}$ is an acyclic partition of X.*

Proof We need to prove that for all $p, p' \in \hat{P}$ one has the equivalence

$$p \preceq_{\mathcal{E}_{\mathcal{M}}} p' \quad \Leftrightarrow \quad p \preccurlyeq p'. \tag{7.10}$$

First, we will prove that

$$E_p \cap \operatorname{cl} E_{p'} \neq \varnothing \quad \Rightarrow \quad p \preccurlyeq p'. \tag{7.11}$$

Let $x \in E_p \cap \operatorname{cl} E_{p'}$ be arbitrary. Then $x \in \operatorname{cl} y$ for a $y \in E_{p'}$. Thus, $\sigma := y \cdot x$ is a path. Moreover, one has $\sigma^{\sqsubset} = y \in E_{p'}$ and $\sigma^{\sqsupset} = x \in E_p$, which proves (7.11). Since, by Proposition 7.2.10, the relation $\preccurlyeq$ is transitive, it follows from property (7.11) that the left-hand side of (7.10) implies the right-hand side of (7.10).

For the reverse implication, assume that $p \preccurlyeq p'$. Let $\varrho = x_0 \cdot x_1 \cdot \ldots \cdot x_n$ be a path from $E_{p'}$ to E_p. Since $\mathcal{E}_{\mathcal{M}}$ is a partition, for each $i = 0, 1, 2, \ldots, n$ one can find an index $p_i \in \hat{P}$ such that $x_i \in E_{p_i}$. In particular, $p_0 = p'$ and $p_n = p$. Moreover, since the inclusion $x_{i+1} \in \operatorname{cl} x_i \cup [x_i]$ holds, we see that $E_{p_{i+1}} \cap \operatorname{cl} E_{p_i} \neq \varnothing$ or $E_{p_{i+1}} = E_{p_i}$. It follows that $p_{i+1} \preceq_{\mathcal{E}_{\mathcal{M}}} p_i$, and therefore $p \leq_{\mathcal{E}_{\mathcal{M}}} p'$. Thus, (7.10) is proved. Hence, one obtains from Proposition 7.2.10 that $\leq_{\mathcal{E}_{\mathcal{M}}}$ is a partial order on $\hat{P}$, and this finally implies that $\mathcal{E}_{\mathcal{M}}$ is an acyclic partition. □

7.3 Connection Matrices and Heteroclinics

Proposition 7.2.11 enables us to consider $(\hat{P}, \preccurlyeq)$ as a poset. We also consider it as an object of DPSET with the distinguished subset given by

$$(\hat{P})_\star := P. \tag{7.12}$$

Note that such a definition of $(\hat{P})_\star$ satisfies (4.16), because it follows from Corollary 7.2.8 that every $V \in \mathcal{M}^\perp$ must be regular, which in turn implies that $C(V)$ is inessential. These observations let us define the Conley complex and associated connection matrix of a given Morse or block decomposition as the Conley complex and connection matrix of $\mathcal{E}_{\mathcal{M}}$ (see Definition 6.1.4).

Definition 7.3.1 (Conley Complex and Connection Matrix) The *Conley complex* and the associated *connection matrix* of a Morse or block decomposition $\mathcal{M}$ of a combinatorial multivector field $\mathcal{V}$ on a Lefschetz complex X is defined as the Conley complex and the associated connection matrix of the acyclic partition $\mathcal{E}_{\mathcal{M}}$, i.e., the Conley complex and respective connection matrix of the poset filtered chain complex $(\hat{P}_{\mathcal{M}}, C(X), \partial^\kappa)$.

Example 7.3.2 (Nonuniqueness via Subdivision, ◁ 7.1.2 ▷ 8.4.9) All three combinatorial multivector fields $\mathcal{V}_0$, $\mathcal{V}_1$, and $\mathcal{V}_2$ in Fig. 7.1 have a common Morse decomposition given as the family

$$\{\{\mathbf{A}\}, \{\mathbf{B}\}, \{\mathbf{AB}\}, \{\mathbf{CD}\}, \{\mathbf{CE}\}\},$$

which consists of singletons. However, the acyclic partition $\mathcal{E}_i := \mathcal{E}_{\mathcal{M},\mathcal{V}_i}$ is different for each index $i \in \{0, 1, 2\}$. In these simple examples, the collection $\mathcal{E}_i$ coincides with $\mathcal{V}_i$. The filtered chain complex $(\hat{P}_{\mathcal{M},\mathcal{V}_i}, C(X), \partial^\kappa)$ coincides with the filtered chain complex considered in Example 4.3.3. Therefore, the connection matrices given by (5.2) and (5.29) are the connection matrices of $\mathcal{M}$ for $\mathcal{V}_0$. We know that these matrices are not equivalent (see Example 5.2.8 and Example 5.5.7). ◊

Given a Morse decomposition $\mathcal{M} = (M_p)_{p \in P}$ and a subset $Q \subset P$, we define the set $M(Q)$ as the collection of cells $x \in X$ for which there exists an essential solution ϱ through x and elements $q_-, q_+ \in Q$ such that both $\operatorname{uim}^-(\varrho) \subset M_{q^-}$ and $\operatorname{uim}^+(\varrho) \subset M_{q^+}$ are satisfied.

Proposition 7.3.3 *The above-defined set $M(Q) \subset X$ is an isolated invariant set for $\mathcal{V}$. Moreover, the family $\mathcal{M}_Q := (M_q)_{q \in Q}$ is a Morse decomposition of $M(Q)$.*

Proof The first statement is the subject of [30, Theorem 7.4]. The second statement is straightforward. □

As an immediate consequence of Theorem 5.4.1 and Corollary 5.4.2, we obtain the following proposition.

Proposition 7.3.4 *Assume that $(P, \bar{C}, \bar{d})$ is a Conley complex of a Morse decomposition $\mathcal{M} = (M_p)_{p \in P}$ of an isolated invariant set S. If $Q \subset P$ is convex with respect to the partial order $\leq$ in P, then $(Q, \bar{C}_Q, \bar{d}_{QQ})$ is a Conley complex of $\mathcal{M}_Q$ considered as a Morse decomposition of $M(Q)$. Moreover, the Conley index of $M(Q)$ is isomorphic to the homology module $H(\bar{C}_Q, \bar{d}_{QQ})$.* □

Clearly, every singleton $\{p\} \subset P$ is convex in P. Since, according to the definition of the Conley complex, one always has $\bar{d}_{pp} = 0$, we immediately get the following corollary from Proposition 7.3.3.

Corollary 7.3.5 *Consider again a Conley complex $(P, \bar{C}, \bar{d})$ of a Morse decomposition $\mathcal{M} = (M_p)_{p \in P}$ of an isolated invariant set S of $\mathcal{V}$. Then for every $p \in P$, the Conley index $\operatorname{Con}(M_p)$ of the Morse set M_p is isomorphic to $\bar{C}_p$.* □

In terms of applications, one of the most important results of classical connection matrix theory states that if the collection $\{M_p\}_{p \in P}$ is a Morse decomposition indexed by a poset P, if the elements $p, q \in P$ are such that q covers p, and if the entry in the q-th row and p-th column of the connection matrix is nonzero, then there must be a heteroclinic connection running from M_q to M_p. We have an analogous result for combinatorial multivector fields.

Theorem 7.3.6 (Existence of Heteroclinics) *Let $(P, \bar{C}, \bar{d})$ be a Conley complex of a Morse decomposition $\mathcal{M} = (M_p)_{p \in P}$, and let A denote the associated connection matrix. Then A is a block matrix $(A_{rs})_{r,s \in P}$.*

Consider two poset elements $p, q \in P$ such that $p < q$ and that the set $\{p, q\}$ is convex in P. If $A_{pq} \neq 0$, then there is a heteroclinic connection from M_q to M_p, that is, an essential solution ϱ satisfying

$$\operatorname{uim}^-(\varrho) \subset M_q \quad \textit{and} \quad \operatorname{uim}^+(\varrho) \subset M_p. \tag{7.13}$$

Proof Assume to the contrary that there is no heteroclinic connection from M_q to M_p. Set $M_{pq} := M(\{p, q\})$. Then it follows from Proposition 7.3.3 that M_{pq} is an isolated invariant set with Morse decomposition $\{M_p, M_q\}$. We will prove that

$$M_{pq} = M_p \cup M_q. \tag{7.14}$$

For this, consider an arbitrary element $x \in M_{pq}$. By the definition of the set M_{pq}, there are $M_-, M_+ \in \{M_p, M_q\}$ and an essential solution ϱ through x in M_{pq} such that both $\operatorname{uim}^-(\varrho) \subset M^-$ and $\operatorname{uim}^+(\varrho) \subset M^+$ are satisfied. We cannot have $M_- = M_p$ and $M_+ = M_q$, because then $p > q$, which contradicts the assumption $p < q$. We also cannot have $M_- = M_q$ and $M_+ = M_p$, since otherwise ϱ would be a heteroclinic connection from M_q to M_p. Hence, we either have the equalities $M_- = M_+ = M_p$, or alternatively $M_- = M_+ = M_q$. Since $\{M_p, M_q\}$ is a Morse decomposition of M_{pq}, in the first case, we get $x \in \operatorname{im}\gamma \subset M_p$, and in the second case, we get $x \in \operatorname{im}\gamma \subset M_q$. This proves that $M_{pq} \subset M_p \cup M_q$. Since the opposite inclusion is obvious, the proof of the equality (7.14) is complete. We also have

$$M_q \cap \operatorname{cl} M_p = \varnothing, \tag{7.15}$$

$$M_p \cap \operatorname{cl} M_q = \varnothing. \tag{7.16}$$

Indeed, to prove (7.15), assume to the contrary that $M_q \cap \operatorname{cl} M_p \neq \varnothing$. Then $\partial^\kappa_{M_q M_p} \neq 0$, which gives $q < p$, because ∂^κ is filtered. This contradicts our assumption $p < q$. To see (7.16), assume to the contrary that there exists an $x \in M_p \cap \operatorname{cl} M_q$. Then one has $x \in \operatorname{cl} y$ for some $y \in M_q$, and in view of $M_p \neq \varnothing \neq M_q$, we can find an essential solution γ_q through y in M_q and γ_p through x in M_p. But then $\gamma_q^- \cdot \gamma_p^+$ is a heteroclinic connection from M_q to M_p, a contradiction.

It follows from the identities (7.14)–(7.16) that M_{pq}, M_p, and M_q satisfy the assumptions of [30, Theorem 5.19]. This allows us to conclude that

$$\operatorname{Con}(M_{pq}) = \operatorname{Con}(M_p) \oplus \operatorname{Con}(M_q)$$

and, in consequence, that

$$n_{pq} = n_p + n_q, \tag{7.17}$$

where n_{pq}, n_p, and n_q denote, respectively, the dimensions of $\operatorname{Con}(M_{pq})$, $\operatorname{Con}(M_p)$, and $\operatorname{Con}(M_q)$. However, from Proposition 7.3.4, we know that the block matrix

$$\bar{A}_{pq} := \begin{bmatrix} 0 & A_{pq} \\ 0 & 0 \end{bmatrix}$$

is a connection matrix of the Conley complex for the Morse decomposition $\{M_p, M_q\}$ of M_{pq}, and $\operatorname{Con}(M_{pq})$ is isomorphic to the homology of the Conley complex of the Morse decomposition $\mathcal{M}$ restricted to $\{M_p, M_q\}$. By Corollary 7.3.5, we know that the 2×2 block matrix $\bar{A}_{pq}$ has exactly n_p columns in the first block and n_q columns in the second block. Since $A_{pq} \neq 0$, we see that the kernel of $\bar{A}_{pq}$ is of dimension less than $n_p + n_q$. Therefore, the dimension of the homology of the restricted Conley complex and, in consequence, also the dimension of $\operatorname{Con}(M_{pq})$ is less than $n_p + n_q$, which contradicts (7.17) and proves the theorem. □

Chapter 8
Connection Matrices for Gradient Vector Fields

This last chapter of the book is devoted to the study of the special case of Forman's combinatorial gradient vector fields on regular Lefschetz complexes. In this situation, one can show that the associated connection matrix is uniquely determined and in fact can be determined in a direct way. This is accomplished in a number of steps. We begin by recalling basic properties of combinatorial vectors in the sense of Forman, before we present the definition and basic properties of his concept of combinatorial flow. After discussing the long-term limit of the latter, we can finally explain how it can be used to obtain the unique connection matrix in this setting. The uniqueness part of the assertion will rely heavily on the singleton partition which has already been discussed in Sect. 6.2.

We would like to point out that throughout this section, we consider a very specific class of multivector fields, namely combinatorial gradient vector fields and the Morse decompositions consisting of critical vectors. In this specific setting, we will be able to establish both the existence of the Conley complex and the uniqueness of the associated connection matrix directly, without reference to Theorem 5.3.2. Thus, we will obtain results which are valid on arbitrary regular Lefschetz complexes.

8.1 Forman's Combinatorial Flow

In the following, we assume that $\mathcal{V}$ is a combinatorial gradient vector field in the sense of Forman on a Lefschetz complex X. In terms of its definition, we say that a combinatorial multivector W is a *combinatorial vector* if its cardinality satisfies $\operatorname{card} W \leq 2$. A combinatorial multivector field whose multivectors are just vectors is then called a *combinatorial vector field*, a concept originally introduced by Forman [18, 19] in a slightly different but equivalent form. In the case of a combinatorial vector field, the critical vectors are precisely the singletons, and the

M. Mrozek, T. Wanner, *Connection Matrices in Combinatorial Topological Dynamics*, SpringerBriefs in Mathematics,
https://doi.org/10.1007/978-3-031-87600-4_8

regular vectors are precisely the doubletons, i.e., vectors of cardinality two (see Proposition 3.5.9). Therefore, every gradient-like combinatorial vector field is a gradient combinatorial vector field.

Let $C \subset \mathcal{V}$ denote the collection of critical vectors in $\mathcal{V}$. Since we assume that $\mathcal{V}$ is a combinatorial vector field, it follows from Proposition 3.5.9 that C is exactly the collection of singletons in $\mathcal{V}$. Moreover, we know from Proposition 7.2.4 that C is a Morse decomposition of $\mathcal{V}$ self-indexed by the poset $P = C$ with partial order $\leq_{\mathcal{V}}$ restricted to C. Then, clearly, the partition $\mathcal{E}_C$ induced by C is $\mathcal{V}$. Also, since we take $(\hat{P})_\star := P$ (see (7.12)) in the construction of the Conley complex, we have the equality $\mathcal{V}_\star = C$. Therefore, one obtains the following straightforward proposition.

Proposition 8.1.1 *Assume that $\mathcal{V}$ is a gradient combinatorial vector field. Then the collection C is a Morse decomposition of $\mathcal{V}$, the induced acyclic partition of the Lefschetz complex X is given by $\mathcal{E}_C = \mathcal{V}$, and the Conley complex of C is the Conley complex of the filtered chain complex $(\mathcal{V}, C(X), \partial^\kappa)$ with $\mathcal{V}_\star = C$.* □

The aim of this chapter is to prove that the Morse decomposition C has precisely one connection matrix and that this connection matrix coincides with the matrix of the boundary operator of the associated Morse complex, see [19, Section 7]. In order to make this statement precise, we first need to develop some concepts, with the current section focusing on properties of combinatorial vectors.

Unlike a general multivector, a combinatorial vector $W \in \mathcal{V}$ contains a unique minimal element and a unique maximal element in W with respect to the face relation $\leq_\kappa$. We denote them by W^- and W^+, respectively, and we extend this notation to cells by writing $x^- := [x]^-$ and $x^+ := [x]^+$. Note that a combinatorial vector is given by $W = \{W^-, W^+\}$, and W is critical if and only if we have $W^+ = W^-$, which in turn is equivalent to assuming that W is a singleton. As a consequence of Proposition 3.5.4, one obtains for a vector W the inequality

$$0 \leq \dim W^+ - \dim W^- \leq 1, \tag{8.1}$$

and for all $x \in W$, one has

$$\dim W^- \leq \dim x \leq \dim W^+. \tag{8.2}$$

We say that a cell x is a *tail* if $x = x^- \neq x^+$ and a *head* if $x = x^+ \neq x^-$. Clearly, if a cell is neither a tail nor a head, then it is critical. We denote the subsets of critical cells, tails, and heads by $X^c(\mathcal{V})$, $X^-(\mathcal{V})$, and $X^+(\mathcal{V})$, respectively, and shorten this notation to X^c, X^-, and X^+ whenever $\mathcal{V}$ is clear from context. Note that the collection C of critical vectors of $\mathcal{V}$ is precisely $\{\{x\} \mid x \in X^c\}$.

For a combinatorial vector $V \in \mathcal{V}$, we set $\dim V := \dim V^+$. Moreover, recall that $\mathcal{V}$, as a gradient vector field, is an acyclic partition of X according to Proposition 7.2.2. In particular, the collection $\mathcal{V}$ is a poset with the partial order $\leq_{\mathcal{V}}$. Then one has the following proposition.

Proposition 8.1.2 *Assume that $\mathcal{V}$ is a gradient vector field on a Lefschetz complex X. Then the following hold:*

(i) The map $\dim : (\mathcal{V}, \leq_{\mathcal{V}}) \to (\mathbb{N}_0, \leq)$ *is order preserving.*
(ii) If $V <_{\mathcal{V}} W$ and V is critical, then $\dim V < \dim W$.

Proof Since, by definition, the relation $\leq_{\mathcal{V}}$ is the transitive closure of the relation $\preceq_{\mathcal{V}}$, to prove (i), it suffices to verify that the inequality $V \cap \operatorname{cl} W \neq \varnothing$ for $V, W \in \mathcal{V}$ implies $\dim V \leq \dim W$. Thus, assume that $V \cap \operatorname{cl} W \neq \varnothing$. Let $x \in V \cap \operatorname{cl} W$. If $x \in W$, then $V \cap W \neq \varnothing$, which implies $V = W$, because $\mathcal{V}$ is a partition. In particular, this gives $\dim V = \dim W$. Thus, consider now the case that $x \notin W$. Since we have the inclusion $W^- \in \operatorname{cl} W^+$, one obtains $\operatorname{cl} W = \operatorname{cl}\{W^-, W^+\} = \operatorname{cl} W^- \cup \operatorname{cl} W^+ = \operatorname{cl} W^+$. It follows that $x \in \operatorname{cl} W^+ \setminus \{W^+\}$, and therefore $\dim x < \dim W^+$. Finally, since we assumed that $x \in V$, we get from (8.1) and (8.2) that $\dim x \geq \dim V^- \geq \dim V^+ - 1 = \dim V - 1$. Thus, we finally get $\dim V \leq \dim x + 1 \leq \dim W^+ = \dim W$.

To see (ii), without loss of generality, we may assume that $V \prec_{\mathcal{V}} W$, that is, $V \neq W$ and $V \cap \operatorname{cl} W \neq \varnothing$. Observe that since V is critical, both $V = \{x\}$ and $\dim V = \dim x$ are satisfied. In view of $x \notin W$ and $x \in \operatorname{cl} W = \operatorname{cl} W^+$, we have $\dim x < \dim W^+$. Therefore, $\dim V = \dim x < \dim W^+ = \dim W$. □

The next piece of the puzzle is Forman's combinatorial flow, which he introduced in [19, Definition 6.2] and which we use in a slightly more general context. More precisely, we now consider a regular Lefschetz complex X. Then with a combinatorial gradient vector field $\mathcal{V}$ on X, one can associate the map $\Gamma_{\mathcal{V}} : C(X) \to C(X)$ of degree $+1$, which for $x \in X$ is defined by

$$\Gamma_{\mathcal{V}} x := \begin{cases} 0 & \text{if } x^- = x^+, \\ -\kappa(x^+, x^-)^{-1} \langle x, x^- \rangle x^+ & \text{otherwise.} \end{cases} \tag{8.3}$$

In the sequel, we drop the subscript in $\Gamma_{\mathcal{V}}$ whenever $\mathcal{V}$ is clear from context. Then the following propositions are straightforward.

Proposition 8.1.3 *If X is a regular Lefschetz complex and $\mathcal{V}$ is a combinatorial gradient vector field on X, then for every $c \in C(X)$ we have $|\Gamma c| \subset X^+$.* □

Proposition 8.1.4 *Suppose that X is a regular Lefschetz complex and $\mathcal{V}$ is a combinatorial gradient vector field on X. Then for all $x, u \in X$, we have both the implication*

$$\langle u, \Gamma \partial x \rangle \neq 0 \quad \Rightarrow \quad u \in X^+ \text{ and } \langle u, \Gamma \partial x \rangle = -\kappa(x, u^-)\kappa(u^+, u^-)^{-1} \tag{8.4}$$

and the implication

$$\langle u, \partial \Gamma x \rangle \neq 0 \quad \Rightarrow \quad x \in X^- \text{ and } \langle u, \partial \Gamma x \rangle = -\kappa(x^+, x^-)^{-1} \langle u, \partial x^+ \rangle. \tag{8.5}$$

Proof The definition of the boundary operator of a Lefschetz complex given in (3.9) implies

$$\Gamma\partial x = \Gamma\left(\sum_{y\in X}\kappa(x,y)y\right) = \sum_{y\in X}\kappa(x,y)\Gamma y$$
$$= -\sum_{y\in X,\,\Gamma y\neq 0}\kappa(x,y)\kappa(y^+,y^-)^{-1}\langle y,y^-\rangle y^+.$$

Hence, we have

$$\langle u,\Gamma\partial x\rangle = -\sum_{y\in X,\,\Gamma y\neq 0}\kappa(x,y)\kappa(y^+,y^-)^{-1}\langle y,y^-\rangle\langle u,y^+\rangle.$$

Thus, if the inequality $\langle u,\Gamma\partial x\rangle \neq 0$ holds, then there is a $y\in X$ such that $y^-\neq y^+$ and that all factors in $\kappa(x,y)\langle y,y^-\rangle\langle u,y^+\rangle$ are nonzero. In particular, $\langle y,y^-\rangle\neq 0$ implies $y=y^-$. Hence, one obtains $y\in X^-$, and from $\langle u,y^+\rangle\neq 0$, we further get both $u=y^+\in X^+$ and $y=y^-=[y^+]^-=[u]^-=u^-$. It follows that such a cell y is unique and $\langle u,\Gamma\partial x\rangle = -\kappa(x,y)\kappa(y^+,y^-)^{-1} = -\kappa(x,u^-)\kappa(u^+,u^-)^{-1}$. This proves the implication (8.4).

To see (8.5), observe that if $\langle u,\partial\Gamma x\rangle\neq 0$, then $\Gamma x\neq 0$. Hence, $x^-\neq x^+$, and one further obtains

$$\langle u,\partial\Gamma x\rangle = \langle u,-\kappa(x^+,x^-)^{-1}\langle x,x^-\rangle\partial x^+\rangle = -\kappa(x^+,x^-)^{-1}\langle x,x^-\rangle\langle u,\partial x^+\rangle.$$

Therefore, if $\langle u,\partial\Gamma x\rangle\neq 0$, then $\langle x,x^-\rangle\neq 0\neq\langle u,\partial x^+\rangle$. In particular, we get $x=x^-$, which in turn means that $x\in X^-$ and $\langle u,\partial\Gamma x\rangle = -\kappa(x^+,x^-)^{-1}\langle u,\partial x^+\rangle$. This identity finally proves (8.5). □

For an arbitrary combinatorial gradient vector field $\mathcal{V}$ on any regular Lefschetz complex X, we follow Forman [19, Theorem 6.4] and define the associated *combinatorial flow* $\Phi := \Phi_{\mathcal{V}} : C(X)\to C(X)$ on generators $x\in X$ through the formula

$$\Phi_{\mathcal{V}}x := x + \partial\Gamma_{\mathcal{V}}x + \Gamma_{\mathcal{V}}\partial x. \tag{8.6}$$

In the sequel, whenever $\mathcal{V}$ is clear from the context, we drop the subscript in $\Phi_{\mathcal{V}}$. Notice that the map Φ is a degree zero module homomorphism which satisfies the identity $\Phi\partial = \partial + \partial\Gamma\partial = \partial\Phi$. Hence, the map Φ is in fact a chain map. The definition of the combinatorial flow is illustrated in the following example.

Example 8.1.5 (Three Forman Gradient Vector Fields ◁ 2.6.3 ▷ 8.2.4) We return to the setting of Example 2.6.3, where we introduced the three combinatorial vector fields shown in Fig. 2.6. For this, consider the underlying simplicial com-

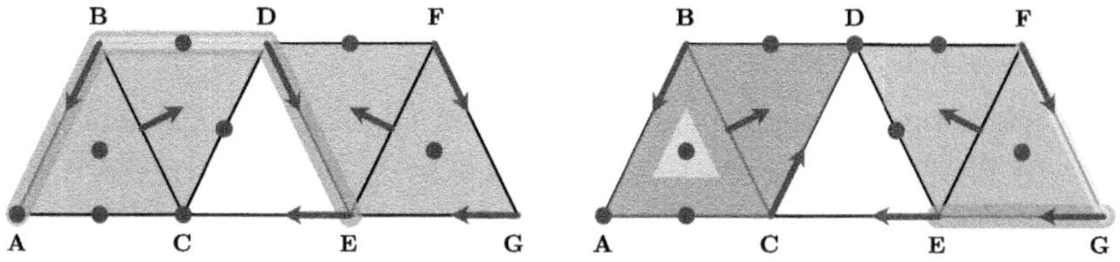

Fig. 8.1 *The combinatorial flow* Φ. The above two panels illustrate the action of the combinatorial flows associated with the two combinatorial vector fields introduced in the top two rows of Fig. 2.6. In the left panel, both the chain **BD** and its image $\Phi_{\mathcal{V}_1}(\mathbf{BD})$ are shown in yellow and blue, respectively. In the panel on the right, one can find $\Phi_{\mathcal{V}_2}(\mathbf{ABC})$ and $\Phi_{\mathcal{V}_2}(\mathbf{FG})$, where again the arguments are depicted in yellow and the images under the combinatorial flow in blue

plex X as a Lefschetz complex with $\mathbb{Z}_2$-coefficients, that is, the simplices making up the complex X are not oriented.

In order to describe the action of the combinatorial flows Φ associated with the two gradient combinatorial vector fields shown in Fig. 8.1, one first needs to understand the map Γ defined in (8.3). It can immediately be seen that due to the employed $\mathbb{Z}_2$-coefficients, the image $\Gamma(x)$ is nonzero if and only if $x = x^- \neq x^+$, and in this case, we have $\Gamma(x^-) = x^+$.

With this, we can now turn our attention to a few sample combinatorial flow computations. We begin by considering the combinatorial vector field $\mathcal{V}_1$ shown in the left panel of Fig. 8.1. For the edge **BD**, one obtains

$$\begin{aligned}\Phi_{\mathcal{V}_1}(\mathbf{BD}) &= \mathbf{BD} + \partial(\Gamma_{\mathcal{V}_1}(\mathbf{BD})) + \Gamma_{\mathcal{V}_1}(\partial\mathbf{BD}) \\ &= \mathbf{BD} + \partial(0) + \Gamma_{\mathcal{V}_1}(\mathbf{B}+\mathbf{D}) \quad = \quad \mathbf{BD}+\mathbf{AB}+\mathbf{DE},\end{aligned}$$

and this is illustrated in the left image of Fig. 8.1. The edge **BD** is shown in yellow, while the chain $\Phi_{\mathcal{V}_1}(\mathbf{BD})$ is drawn in blue. Notice that the action of the combinatorial flow $\Phi_{\mathcal{V}_1}$ encodes potential solution paths of the combinatorial vector field $\mathcal{V}_1$ starting at the edge **BD**.

For the gradient combinatorial vector field $\mathcal{V}_2$ shown in the right panel of Fig. 8.1, we consider two examples. First, the edge **FG** has the image

$$\Phi_{\mathcal{V}_2}(\mathbf{FG}) = \mathbf{FG} + \partial(0) + \Gamma(\mathbf{F}+\mathbf{G}) = \mathbf{FG}+\mathbf{FG}+\mathbf{EG} = \mathbf{EG},$$

since both $\Gamma(\mathbf{FG}) = 0$ and $\mathbf{FG}+\mathbf{FG} = 0$ are satisfied. In addition, for the two-dimensional cell **ABC**, one obtains the identity

$$\begin{aligned}\Phi_{\mathcal{V}_2}(\mathbf{ABC}) &= \mathbf{ABC} + \partial(0) + \Gamma_{\mathcal{V}_2}(\mathbf{AB}+\mathbf{AC}+\mathbf{BC}) \\ &= \mathbf{ABC} + 0 + 0 + \mathbf{BCD} \quad = \quad \mathbf{ABC}+\mathbf{BCD},\end{aligned}$$

in view of $\Gamma_{\mathcal{V}_2}(\mathbf{ABC}) = 0$ and $\Gamma_{\mathcal{V}_2}(\mathbf{AB}) = \Gamma_{\mathcal{V}_2}(\mathbf{AC}) = 0$, as well as using our above discussion to show that $\Gamma_{\mathcal{V}_2}(\mathbf{BC}) = \mathbf{BCD}$. Both of these images under the combinatorial flow $\Phi_{\mathcal{V}_2}$ are depicted in blue in the right panel of Fig. 8.1, while the considered arguments are shown in yellow. $\Diamond$

The above sample computations demonstrate that the combinatorial flow in some sense encodes the potential paths that solutions of the combinatorial gradient vector field can take. For later use, we now present four propositions, which provide more rigorous insight into the action of the combinatorial flow Φ.

Proposition 8.1.6 *Suppose that X is a regular Lefschetz complex and $\mathcal{V}$ is a combinatorial gradient vector field on X. Then for all $x \in X$, we have*

$$\langle x, \Phi x\rangle = \begin{cases} 1 & \text{if } x \in X^c, \\ 0 & \text{otherwise.} \end{cases}$$

Proof Notice that the identity $\langle x, \Phi x\rangle = \langle x, x\rangle + \langle x, \partial\Gamma x\rangle + \langle x, \Gamma\partial x\rangle$ holds. In the case $x \in X^c$, we get from Proposition 8.1.4 that $\langle x, \partial\Gamma x\rangle = 0$ and $\langle x, \Gamma\partial x\rangle = 0$, which immediately implies $\langle x, \Phi x\rangle = \langle x, x\rangle = 1$.

Consider now the case $x \in X^-$. Then one has $x = x^- \notin X^+$, and we further obtain from (8.4) the equality $\langle x, \Gamma\partial x\rangle = 0$. Therefore, the definition (8.3) yields

$$\langle x, \Phi x\rangle = \langle x, x\rangle + \langle x, \partial\Gamma x\rangle = \\ 1 - \kappa(x^+, x^-)^{-1}\langle x, x^-\rangle\langle x, \partial x^+\rangle = 1 - \kappa(x^+, x^-)^{-1}\kappa(x^+, x^-) = 0.$$

Finally, we consider the remaining case $x \in X^+$. Then we have $x \notin X^-$, and the implication (8.5) shows that $\langle x, \partial\Gamma x\rangle = 0$. In view of $\partial x = \sum_{y\in X}\kappa(x, y)y$, this gives

$$\begin{aligned} \langle x, \Phi x\rangle &= \langle x, x\rangle + \langle x, \Gamma\partial x\rangle = 1 + \langle x, \sum_{y\in X}\kappa(x, y)\Gamma y\rangle \\ &= 1 - \langle x, \sum_{y\in X,\ \Gamma y\neq 0}\kappa(x, y)\kappa(y^+, y^-)^{-1}\langle y, y^-\rangle y^+\rangle \\ &= 1 - \sum_{y\in X,\ \Gamma y\neq 0}\kappa(x, y)\kappa(y^+, y^-)^{-1}\langle y, y^-\rangle\langle x, y^+\rangle \\ &= 1 - \kappa(x^+, x^-)\kappa(x^+, x^-)^{-1} = 0, \end{aligned}$$

because the only nonzero term in the last sum occurs for the element $y \in X$ which satisfies both $y^+ = x = x^+$ and $y = y^- = x^-$. $\square$

Proposition 8.1.7 *Suppose that X is a regular Lefschetz complex and $\mathcal{V}$ is a combinatorial gradient vector field on X. Then we have*

$$\Phi(C(X^+)) \subset C(X^+).$$

Proof Since Φ is a homomorphism, it suffices to show that the inclusion $x \in X^+$ implies $|\Phi x| \subset X^+$. Thus, let $x \in X^+$ and $u \in |\Phi x|$. Then $\langle u, \Phi x\rangle \neq 0$. From (8.5), one further obtains $\langle u, \partial\Gamma x\rangle = 0$. Therefore, $0 \neq \langle u, \Phi x\rangle = \langle u, x\rangle + \langle u, \Gamma\partial x\rangle$, which implies $\langle u, x\rangle \neq 0$ or $\langle u, \Gamma\partial x\rangle \neq 0$. If $\langle u, x\rangle \neq 0$, then $u = x \in X^+$. If $\langle u, \Gamma\partial x\rangle \neq 0$ holds, then we get from (8.4) that $u \in X^+$. Hence, $|\Phi x| \subset X^+$. □

Proposition 8.1.8 *Assume that X is a regular Lefschetz complex, that $\mathcal{V}$ is a combinatorial gradient vector field on X, and that $\mathcal{A} \in$ Down$(\mathcal{V}, \leq_{\mathcal{V}})$. Then the set $A := |\mathcal{A}|$ is $\mathcal{V}$-compatible and closed, and we have $\Phi(C(A)) \subset C(A)$. Therefore, the restriction $\Phi_{|C(A)} : (C(A), \partial^{\kappa}_{|C(A)}) \to (C(A), \partial^{\kappa}_{|C(A)})$ is again a well-defined chain map.*

Proof Clearly, the set A is $\mathcal{V}$-compatible, and it follows from Proposition 6.1.2(i) that A is closed. Take an $x \in A$. We will prove that $\Phi x \in C(A)$. Since A is closed, we have $|\partial x| \subset A$. Since A is $\mathcal{V}$-compatible, we have $|\Gamma_{\mathcal{V}} x| \subset A$. Thus, again by closedness and $\mathcal{V}$-compatibility of A, we get both $|\partial\Gamma_{\mathcal{V}} x| \subset A$ and $|\Gamma_{\mathcal{V}}\partial x| \subset A$. This in turn implies

$$|\Phi x| = |x + \partial\Gamma_{\mathcal{V}} x + \Gamma_{\mathcal{V}}\partial x| \subset |x| \cup |\partial\Gamma_{\mathcal{V}} x| \cup |\Gamma_{\mathcal{V}}\partial x| \subset A,$$

which gives $\Phi x \in C(A)$ and completes the proof. □

Proposition 8.1.9 *Suppose that X is a regular Lefschetz complex and that $\mathcal{V}$ is a combinatorial gradient vector field on X. Assume further that $x, y \in X$. Then the following hold:*

(i) If $y \in |\Phi x|$, then $[y] \leq_{\mathcal{V}} [x]$, and there exists a path from x to y with respect to the multivalued map $\Pi_{\mathcal{V}}$ defined in (7.1).
(ii) If $y \in |\Phi x|$ and $[y] = [x]$, then $y = x$.
(iii) If $x \in |\Phi x|$, then $\langle x, \Phi x\rangle = 1$.

In addition, for any chain $c \in C(X)$, the following holds:

(iv) If $c = \Phi(c)$ and $c \neq 0$, then $|c|$ has to contain a critical cell.

Proof For (i), note that $y \in |\Phi x|$ gives $0 \neq \langle y, \Phi x\rangle = \langle y, x\rangle + \langle y, \partial\Gamma x\rangle + \langle y, \Gamma\partial x\rangle$. Thus, one has either $\langle y, x\rangle \neq 0$, or $\langle y, \partial\Gamma x\rangle \neq 0$, or $\langle y, \Gamma\partial x\rangle \neq 0$. The first case implies $x = y$ and $[x] = [y]$. Consider now the second case $\langle y, \partial\Gamma x\rangle \neq 0$. This inequality yields both $\Gamma x \neq 0$ and $y \in |\partial\Gamma x|$. Hence, we have $x = x^-$ and $\Gamma x = \lambda x^+$ with $\lambda = -\kappa(x^+, x^-)^{-1} \neq 0$. This gives $|\partial\Gamma x| = |\partial x^+|$, as well as $y \in |\partial x^+| \subset \mathrm{cl}\, x^+$. It follows that $y \in [y] \cap \mathrm{cl}[x^+] = [y] \cap \mathrm{cl}[x^-]$, and therefore $[y] \leq_{\mathcal{V}} [x]$. Finally, we consider the third inequality $\langle y, \Gamma \partial x \rangle \neq 0$. Since $\partial x = \sum_{u \in X} \kappa(x, u) u$, we have

$$0 \neq \langle y, \Gamma \partial x \rangle = \sum_{u \in X} \kappa(x, u) \langle y, \Gamma u \rangle,$$

which implies $\langle y, \Gamma u \rangle \neq 0$ for some $u \in |\partial x|$. But then one has $u = u^-$ and $\Gamma u = \lambda u^+$ with $\lambda = -\kappa(u^+, u^-)^{-1} \neq 0$. Hence, the identities $y = u^+$ and $[y] = [u^+] = [u^-]$ are satisfied, as well as the inclusion $u^- \in |\partial x| \subset \mathrm{cl}\, x$. Thus, $[y] \cap \mathrm{cl}[x] \neq \varnothing$, which implies $[y] \leq_{\mathcal{V}} [x]$. Furthermore, in all of these three cases, one can easily construct a path from x to y with respect to the multivalued map $\Pi_{\mathcal{V}}$. This completes the proof of (i).

In order to prove (ii), we observe that $y \in |\Phi x|$ implies $\dim x = \dim y$, since Φ is a degree zero module homomorphism. Since $\mathcal{V}$ is a combinatorial vector field, we further have $\{x^-, x^+\} = [x] = [y] = \{y^-, y^+\}$. Thus, since $\dim x = \dim y$, we must have $x = x^- = y^- = y$ or $x = x^+ = y^+ = y$, which in turn establishes (ii).

Property (iii) is immediate from the definition of chain support.

Finally, consider (iv). Let $y \in |c|$ be such that $[y]$ is a maximal element with respect to $\leq_{\mathcal{V}}$. In other words, there is no $z \in |c|$ for which $[y] <_{\mathcal{V}} [z]$. Since

$$y \in |c| = |\Phi(c)| \subset \bigcup_{x \in |c|} |\Phi(x)|,$$

there exists an $x \in |c|$ with $y \in |\Phi(x)|$. Then part (i) implies $[y] \leq_{\mathcal{V}} [x]$, and in view of our choice of y, this yields $[y] = [x]$. But then (ii) furnishes $y = x$, and an application of (iii) gives $\langle x, \Phi x \rangle = 1$. Together with Proposition 8.1.6, one finally obtains $y = x \in X^c$ as claimed. □

8.2 The Stabilized Combinatorial Flow

As mentioned earlier, the combinatorial flow describes on the level of chains the possible future evolutions of solutions of the underlying combinatorial vector field. Thus, in the case of combinatorial gradient vector fields, iterations of Φ should eventually stabilize and provide intuition on potential connecting orbits between critical cells. In other words, such a stabilization of the combinatorial flow should encode both the Conley complex and the associated connection matrix.

However, before we can show that such a stabilization exists, we first have to study the behavior of iterates of the combinatorial flow. We begin with a result that is an easy application of Proposition 8.1.9 and shows that cells in $|\Phi^n x|$ have to belong to vectors below $[x]$ in the partial order $\leq_{\mathcal{V}}$, regardless of the iteration number $n \in \mathbb{N}$.

Proposition 8.2.1 *Assume that $\mathcal{V}$ is a combinatorial gradient vector field on a regular Lefschetz complex X. If the inclusion $y \in |\Phi^n x|$ holds for some $n \in \mathbb{N}$, then the inequality $[y] \leq_{\mathcal{V}} [x]$ is satisfied.*

Proof We proceed by induction on the natural number n. For $n = 1$, the conclusion is the statement of Proposition 8.1.9. Thus, fix an $n \in \mathbb{N}$ and assume that the conclusion holds for all integers $k \in \{1, 2, \ldots n\}$. Let $y \in |\Phi^{n+1}x|$ be arbitrary, and let $c := \Phi^n x$. Then we have $c = \sum_{w \in |c|} \langle c, w \rangle w$ and $\Phi c = \sum_{w \in |c|} \langle c, w \rangle \Phi w$. Since

$$y \in |\Phi^{n+1}x| = |\Phi c| \subset \bigcup_{w \in |c|} |\Phi w|,$$

one further obtains $y \in |\Phi w|$ for some $w \in |c| = |\Phi^n x|$. Thus, by our induction assumption, this implies $[y] \leq_{\mathcal{V}} [w] \leq_{\mathcal{V}} [x]$. □

As our next step, we study in more detail the specific form of the image chain $\Phi^n x$ if the cell x is critical and $n \in \mathbb{N}$. This result will prove to be essential for identifying the Conley complex associated with $\mathcal{V}$.

Proposition 8.2.2 *Assume that $\mathcal{V}$ is a combinatorial gradient vector field on a regular Lefschetz complex X. Then for every critical cell $x \in X^c$ and every positive integer $n \in \mathbb{N}$, we have the representation*

$$\Phi^n x = x + r_{x,n}, \tag{8.7}$$

where $r_{x,n}$ is a chain satisfying

$$|r_{x,n}| \subset X^+ \cap |\,[x]^{<_{\mathcal{V}}}\,|. \tag{8.8}$$

Proof We proceed again by induction on the natural number n. Assume first that one has $n = 1$ and set $r_{x,1} := \Gamma \partial x$. From Proposition 8.1.3, we see that $|r_{x,1}| \subset X^+$, and, since we assumed $x \in X^c$, one obtains further $\langle x, r_{x,1} \rangle = 0$. The inclusion $x \in X^c$ also shows that $\Gamma x = 0$, and therefore the identity $\Phi x = x + \Gamma \partial x = x + r_{x,1}$ holds. In order to show that $|r_{x,1}| \subset |\,[x]^{<_{\mathcal{V}}}\,|$, take an arbitrary $y \in |r_{x,1}|$. Then we have $\langle y, r_{x,1} \rangle \neq 0$, and since $\langle x, r_{x,1} \rangle = 0$, we get $x \neq y$, that is, $\langle y, x \rangle = 0$. It follows that $\langle y, \Phi x \rangle = \langle y, x \rangle + \langle y, r_{x,1} \rangle = \langle y, r_{x,1} \rangle \neq 0$. Hence, $y \in |\Phi x|$. Thus, Proposition 8.1.9(i) gives $[y] \leq_{\mathcal{V}} [x]$ and $[y] \in [x]^{\leq_{\mathcal{V}}}$. Since $y \neq x$, and in view of the fact that Φ is a degree 0 map, one further has $\dim y = \dim x$, and we also get both $[y] \neq [x]$ and $[y] \in [x]^{<_{\mathcal{V}}}$. This in turn yields the inclusion $y \in |\,[x]^{<_{\mathcal{V}}}\,|$. Therefore, $|r_{x,1}| \subset |\,[x]^{<_{\mathcal{V}}}\,|$, and the proof of (8.7) for $n = 1$ is complete.

Next, fix an integer $k \in \mathbb{N}$, and assume that the equality (8.7) holds for all $n \leq k$ with $r_{x,n}$ satisfying (8.8). Then $\Phi^{k+1}x = \Phi(\Phi^k x) = \Phi x + \Phi r_{x,k} = x + r_{x,1} + \Phi r_{x,k}$, and we set $r_{x,k+1} := r_{x,1} + \Phi r_{x,k}$. It follows from the induction assumption and Proposition 8.1.7 that $|r_{x,k+1}| \subset |r_{x,1}| \cup |\Phi r_{x,k}| \subset X^+$. Since $|r_{x,1}| \subset |\,[x]^{<_{\mathcal{V}}}\,|$, it now suffices to prove that $|\Phi r_{x,k}| \subset |\,[x]^{<_\mathcal{V}}\,|$ in order to see that $|r_{x,k+1}| \subset |\,[x]^{<_\mathcal{V}}\,|$. We will first show that

$$y \in |r_{x,k}| \quad \Rightarrow \quad |\,[y]^{\leq_\mathcal{V}}\,| \subset |\,[x]^{<_\mathcal{V}}\,|. \tag{8.9}$$

For this, let $y \in |r_{x,k}|$. Then the induction assumption yields $y \in |\,[x]^{<_\mathcal{V}}\,|$. Since the set $|\,[x]^{<_\mathcal{V}}\,|$ is $\mathcal{V}$-compatible, it follows that $[y] \in [x]^{<_\mathcal{V}}$. Since $[x]^{<_\mathcal{V}}$ is a down set, one obtains $[y]^{\leq_\mathcal{V}} \subset [x]^{<_\mathcal{V}}$ and $|\,[y]^{\leq_\mathcal{V}}\,| \subset |\,[x]^{<_\mathcal{V}}\,|$. Thus, we proved (8.9). We have $r_{x,k} = \sum_{y\in|r_{x,k}|}\langle r_{x,k}, y\rangle y$ and $\Phi r_{x,k} = \sum_{y\in|r_{x,k}|}\langle r_{x,k}, y\rangle \Phi y$. In view of Proposition 8.1.9(i) and property (8.9), one then obtains the inclusion

$$|\Phi r_{x,k}| \subset \bigcup_{y\in|r_{x,k}|} |\Phi y| \subset \bigcup_{y\in|r_{x,k}|} |\,[y]^{\leq_\mathcal{V}}| \subset |\,[x]^{<_\mathcal{V}}\,|,$$

which completes the induction argument, and therefore the proof. □

The statement of the above proposition can be illustrated using the sample computations in Example 8.1.5, two of which have arguments which are critical cells. As shown in the left panel of Fig. 8.1, for the gradient combinatorial vector field $\mathcal{V}_1$, one has the identity $\Phi_{\mathcal{V}_1}(\mathbf{BD}) = \mathbf{BD} + \mathbf{AB} + \mathbf{DE}$, i.e., in this situation $r_{\mathbf{BD},1} = \mathbf{AB} + \mathbf{DE}$. Similarly, for the vector field $\mathcal{V}_2$ in the right panel, one has $\Phi_{\mathcal{V}_2}(\mathbf{ABC}) = \mathbf{ABC} + \mathbf{BCD}$, which leads to $r_{\mathbf{ABC},1} = \mathbf{BCD}$. In both cases, the chain $r_{x,1}$ clearly satisfies (8.8).

After these preparations, we can finally show that iterations of the combinatorial flow have to stabilize. In the context of the Morse complex, this was already shown by Forman in [19]. The following result is modeled after [19, Theorem 7.2], yet adapted to our situation.

Proposition 8.2.3 *Suppose that $\mathcal{V}$ denotes a combinatorial gradient vector field on a regular Lefschetz complex X. In addition, we consider the induced combinatorial flow $\Phi = \Phi_\mathcal{V} : C(X) \to C(X)$ as defined in (8.6). Then there exists an $N \in \mathbb{N}$ with*

$$\Phi^n = \Phi^N \quad \textit{for all} \quad n \geq N.$$

We define $\Phi^\infty := \Phi^N$ and call it the stabilized combinatorial flow.

Proof We show that for every cell $x \in X$ there exists an integer $n_x \in \mathbb{N}$ such that $\Phi^{n_x+1}x = \Phi^{n_x}x$ holds. Then it is straightforward to see that choosing N as the maximum of $\{n_x \mid x \in X\}$ satisfies the statement of the proposition.

In order to prove the above claim, we proceed by induction over the number of vectors which are strictly below $[x]$ with respect to $<_\mathcal{V}$, that is, by induction over the number

$$L(x) := \#\{V \in \mathcal{V} \mid V <_\mathcal{V} [x]\}.$$

Suppose first that the element $x \in X$ satisfies $L(x) = 0$. If we have $\Phi x = 0$, then clearly $\Phi^2 x = \Phi x = 0$ and the statement follows. Otherwise there exists a cell $y \in X$ with $y \in |\Phi x|$. But then Proposition 8.1.9(i) implies $[y] \leq_{\mathcal{V}} [x]$, and the assumed identity $L(x) = 0$ further yields $[y] = [x]$. This in turn gives both $y = x$ and $x \in |\Phi x|$, as well as $\langle x, \Phi x \rangle = 1$, in view of Proposition 8.1.9(ii),(iii). In other words, we have to have $\Phi x = x$, and the claim follows again.

Now let $k \in \mathbb{N}$ be arbitrary and assume that we have verified the claim for all elements $z \in X$ with $L(z) < k$. Moreover, let $x \in X$ be a cell with $L(x) = k$. We now distinguish the cases of x being critical or not.

We assume first that x is not a critical cell. Then for all elements $y \in |\Phi x|$, one has to have $[y] <_{\mathcal{V}} [x]$. To show this, suppose otherwise that there is a $y \in |\Phi x|$ with $[y] = [x]$. Then Proposition 8.1.9(ii),(iii) implies $y = x$ and $\langle x, \Phi x \rangle = 1$, and in view of Proposition 8.1.6, this gives $x \in X^c$, a contradiction. Together with Proposition 8.1.9(i), we therefore have $[y] <_{\mathcal{V}} [x]$ for all $y \in |\Phi x|$ as claimed. Thus, every term in the representation

$$\Phi x = \sum_{y \in |\Phi x|} \langle y, \Phi x \rangle y$$

satisfies $L(y) < k$, and according to the inductive hypothesis, there exists, after possibly passing to the maximum of the individual values, a natural number $m \in \mathbb{N}$ such that $\Phi^{m+1} y = \Phi^m y$ for all $y \in |\Phi x|$. This finally furnishes

$$\Phi^{m+2} x = \sum_{y \in |\Phi x|} \langle y, \Phi x \rangle \Phi^{m+1} y = \sum_{y \in |\Phi x|} \langle y, \Phi x \rangle \Phi^m y = \Phi^{m+1} x,$$

and the claim from the beginning of the proof follows for noncritical x.

Finally, suppose that $x \in X^c$. Then according to Proposition 8.2.2, there exists a chain $r_{x,1}$ satisfying (8.8), as well as $\Phi x = x + r_{x,1}$. A straightforward induction argument now implies

$$\Phi^\ell x = x + r_{x,1} + \Phi r_{x,1} + \ldots + \Phi^{\ell-1} r_{x,1} \quad \text{for all} \quad \ell \in \mathbb{N}.$$

Thus, in order to establish that we have the identity $\Phi^{m+1} x = \Phi^m x$ for some natural number $m \in \mathbb{N}$, it suffices to show that $\Phi^m r_{x,1} = 0$. Notice that according to (8.8) we can apply the inductive hypothesis to $r_{x,1}$, and therefore there exists an integer $m \in \mathbb{N}$ such that $\Phi^{m+1} r_{x,1} = \Phi^m r_{x,1}$. This immediately shows that $\Phi^m r_{x,1}$ is a fixed chain under the combinatorial flow Φ. Furthermore, the inclusion in (8.8) and Proposition 8.1.7 yields $|\Phi^m r_{x,1}| \subset X^+$. But then Proposition 8.1.9(iv) implies that we have to have $\Phi^m r_{x,1} = 0$, since the support of $\Phi^m r_{x,1}$ contains no critical cells. This finally establishes the induction step also for a critical cell x, and the proof is complete. □

We would like to point out that in view of this result, the stabilized combinatorial flow Φ^∞ also satisfies the statements in Propositions 8.2.1 and 8.2.2, as long as we replace n by ∞ in their formulations.

Example 8.2.4 (Three Forman Gradient Vector Fields ◁ 8.1.5 ▷ 8.4.7) To illustrate the action of the stabilized flow, we return to Examples 2.6.3 and 8.1.5. For the two gradient combinatorial vector fields $\mathcal{V}_1$ and $\mathcal{V}_2$ shown in the first two rows of Fig. 2.6, and in Fig. 8.1, what are the images of the critical cells under the stabilized combinatorial flow?

We begin by considering the zero-dimensional critical cells in the vector fields. One can immediately see that both their boundaries and their images under Γ are trivial, and therefore we have $\Phi^\infty_{\mathcal{V}_k}(x) = x$ for all of these cells. In addition, for the two-dimensional cells, one easily obtains that $\Phi^\infty_{\mathcal{V}_k}(\mathbf{ABC}) = \mathbf{ABC} + \mathbf{BCD}$, as well as $\Phi^\infty_{\mathcal{V}_k}(\mathbf{EFG}) = \mathbf{EFG} + \mathbf{DEF}$. This follows directly from the computation in Example 8.1.5, together with the fact that $\Phi_{\mathcal{V}_k}(\mathbf{BCD}) = \Phi_{\mathcal{V}_k}(\mathbf{DEF}) = 0$, which in turn implies that the second iterate of $\Phi_{\mathcal{V}_k}$ coincides with the first iterate on critical cells of dimension two.

This leaves us with the one-dimensional critical cells. As it turns out, their images under the stabilized flow can be different with respect to $\mathcal{V}_1$ or $\mathcal{V}_2$. The resulting images $\Phi^\infty_{\mathcal{V}_k}(x)$ are shown in Fig. 8.2 for $k = 1$, and in Fig. 8.3 for $k = 2$. Notice that in almost all of these cases, the image of a one-dimensional cell x under the stabilized flow is given by the support of all connecting orbits between the cell x and zero-dimensional critical cells. The only exception is the image $\Phi^\infty_{\mathcal{V}_1}(\mathbf{DF})$, which lacks the edge **CE**. Yet, since this edge is contained in both connecting orbits between **DF** and **C**, the sum of the associated two chains with respect to $\mathbb{Z}_2$-coefficients leads to its cancelation. ◊

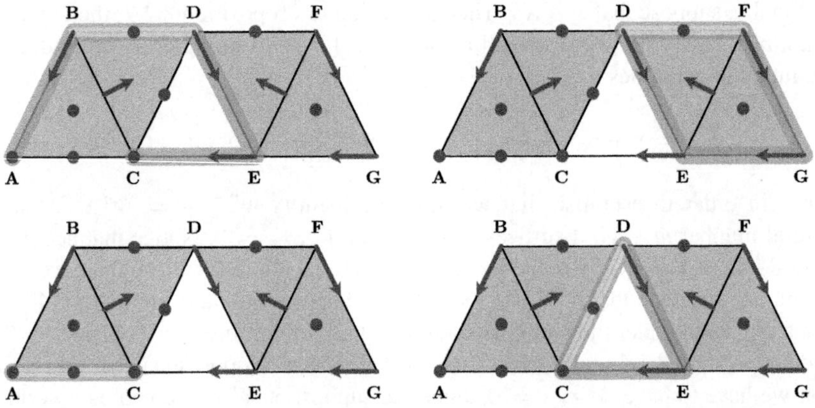

Fig. 8.2 *The stabilized combinatorial flow.* For the gradient combinatorial vector field $\mathcal{V}_1$ introduced in the first row of Fig. 2.6, the four panels show the images $\Phi^\infty_{\mathcal{V}_1}(x)$ for the one-dimensional critical cells x. The chains $\Phi^\infty_{\mathcal{V}_1}(\mathbf{BD})$, $\Phi^\infty_{\mathcal{V}_1}(\mathbf{DF})$, $\Phi^\infty_{\mathcal{V}_1}(\mathbf{AC})$, and $\Phi^\infty_{\mathcal{V}_1}(\mathbf{CD})$ are depicted in blue in the panels from top left to bottom right, respectively

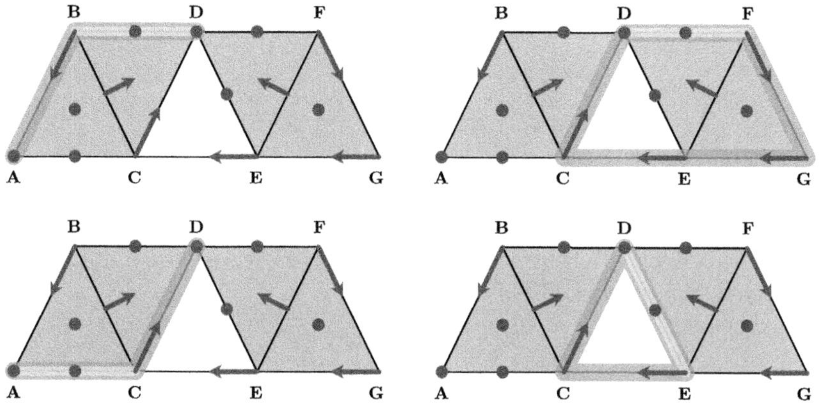

Fig. 8.3 *The stabilized combinatorial flow*. For the gradient vector field $\mathcal{V}_2$ introduced in the second row of Fig. 2.6, the above four panels show the images $\Phi^{\infty}_{\mathcal{V}_2}(x)$ for the one-dimensional critical cells x. From top left to bottom right, they depict the chains $\Phi^{\infty}_{\mathcal{V}_2}(\mathbf{BD})$, $\Phi^{\infty}_{\mathcal{V}_2}(\mathbf{DF})$, $\Phi^{\infty}_{\mathcal{V}_2}(\mathbf{AC})$, and $\Phi^{\infty}_{\mathcal{V}_2}(\mathbf{DE})$, respectively, in blue

8.3 Fixed Chains of the Stabilized Combinatorial Flow

The stabilized combinatorial flow is our main tool for the construction of the Conley complex and its associated connection matrix for a combinatorial gradient vector field $\mathcal{V}$ on a regular Lefschetz complex X. It follows immediately from Proposition 8.2.3 that for every chain $c \in C(X)$, we have

$$\Phi(\Phi^{\infty}(c)) = \Phi(\Phi^{N}(c)) = \Phi^{N+1}(c) = \Phi^{\infty}(c),$$

that is, the image $\Phi^{\infty}(c)$ is a fixed point of the combinatorial flow Φ. We therefore consider the set

$$\operatorname{Fix}\Phi := \{c \in C(X) \mid \Phi c = c\} \tag{8.10}$$

consisting of chains fixed by Φ. One can immediately verify that if $c \in \operatorname{Fix}\Phi$, then $\partial c \in \operatorname{Fix}\Phi$, because $\Phi\partial c = \partial\Phi c = \partial c$. It follows that $(\operatorname{Fix}\Phi, \partial_{|\operatorname{Fix}\Phi})$ is a chain subcomplex of $(C(X), \partial)$. Furthermore, since Φ is a chain map, Proposition 8.2.3 implies that $\Phi^{\infty} : (C(X), \partial) \to (\operatorname{Fix}\Phi, \partial_{|\operatorname{Fix}\Phi})$ is a chain map as well. Yet even more is true, as the following result demonstrates.

Proposition 8.3.1 *Assume that $\mathcal{V}$ is a combinatorial gradient vector field on a regular Lefschetz complex X, and let Φ^{∞} denote the associated stabilized combinatorial flow. Then the restriction $\Phi^{\infty}_{|C(X^c)} : C(X^c) \to \operatorname{Fix}\Phi$ is an isomorphism.*

Moreover, if we define the projection $\Pi : \operatorname{Fix}\Phi \to C(X^c)$ *via*

$$\Pi c := \sum_{x \in X^c} \langle c, x \rangle x \quad \textit{for every} \quad c \in \operatorname{Fix}\Phi, \tag{8.11}$$

then we have $\Phi^\infty \Pi c = c$ *for all* $c \in \operatorname{Fix}\Phi$.

Proof We begin by verifying that $\Phi^\infty_{|C(X^c)}$ is a monomorphism. For this, suppose that $c \in C(X^c)$ is a chain with $\Phi^\infty c = 0$. Then $c = \sum_{x \in X^c} a_x x$, and the representation in equation (8.7) further yields

$$0 = \Phi^\infty c = \sum_{x \in X^c} a_x \Phi^\infty x = c + \sum_{x \in X^c} a_x r_x \in C(X^c) \oplus C(X^+).$$

Hence, it follows that $c = 0$, and therefore $\Phi^\infty_{|C(X^c)}$ is one to one.

In order to prove that $\Phi^\infty_{|C(X^c)} : C(X^c) \to \operatorname{Fix}\Phi$ is onto, let $c \in \operatorname{Fix}\Phi$ be arbitrary and consider the chain $\Pi c \in C(X^c)$ defined in (8.11). Then we have

$$\Phi^\infty \Pi c = \sum_{x \in X^c} \langle c, x \rangle \Phi^\infty x,$$

and for every critical cell $y \in X^c$, the representation of $\Phi^\infty x$ in Proposition 8.2.2 further implies $\langle \Phi^\infty x, y \rangle = \langle x + r_x, y \rangle = \langle x, y \rangle$, see (8.7). Thus for $y \in X^c$, one obtains

$$\langle \Phi^\infty \Pi c, y \rangle = \sum_{x \in X^c} \langle c, x \rangle \langle x, y \rangle = \langle c, y \rangle.$$

Hence, $\langle c - \Phi^\infty \Pi c, y \rangle = 0$ for all $y \in X^c$, i.e., we have $|c - \Phi^\infty \Pi c| \cap X^c = \varnothing$. In view of Proposition 8.1.9(iv) and $c - \Phi^\infty \Pi c \in \operatorname{Fix}\Phi$, this implies the identity $c = \Phi^\infty \Pi c$ and thereby completes the proof of the result. □

The above result shows that the fixed chains under the combinatorial flow Φ are in one-to-one correspondence with chains of critical cells in the Lefschetz complex X. As the following result demonstrates, this fact can be used to both find a suitable basis of $\operatorname{Fix}\Phi$ which is indexed by the critical cells of X, as well as equip this basis with the structure of a Lefschetz complex.

Proposition 8.3.2 *Assume that* $\mathcal{V}$ *is a combinatorial gradient vector field on a regular Lefschetz complex* X*, and recall that* $X^c \subset X$ *denotes the set of all critical cells of* $\mathcal{V}$*. For every* $x \in X^c$*, set*

$$\bar{x} := \Phi^\infty x \in \operatorname{Fix}\Phi.$$

Then the following statements hold:

(i) *The set* $\bar{X} := \{\, \bar{x} \mid x \in X^c \,\}$ *is a basis of* $\operatorname{Fix} \Phi$.

(ii) *For every* $x \in X^c$, *we have the representation*

$$\partial \bar{x} = \sum_{z \in X^c} a_{xz} \bar{z}, \tag{8.12}$$

for uniquely determined coefficients $a_{xz} \in R$.

(iii) *The pair* $(\bar{X}, \bar{\kappa})$, *where the* $\mathbb{Z}$*-gradation on the set* $\bar{X}$ *is induced by the dimension map* $\dim : \bar{X} \ni \bar{x} \mapsto \dim x \in \mathbb{Z}$, *and where* $\bar{\kappa}(\bar{x}, \bar{z})$ *denotes the coefficient* a_{xz} *in equation* (8.12), *is a Lefschetz complex with* $C(\bar{X}) = \operatorname{Fix} \Phi$.

Proof In order to prove (i), consider the projection $\Pi : \operatorname{Fix} \Phi \to C(X^c)$ defined in Proposition 8.3.1, which has been shown to satisfy $\Phi^\infty \Pi = \mathrm{id}_{\operatorname{Fix} \Phi}$. In addition, let $c \in \operatorname{Fix} \Phi$ be arbitrary. Then one has

$$c = \Phi^\infty \Pi c = \Phi^\infty \sum_{x \in X^c} \langle c, x \rangle x = \sum_{x \in X^c} \langle c, x \rangle \Phi^\infty x = \sum_{x \in X^c} \langle c, x \rangle \bar{x},$$

which proves that $\bar{X}$ generates $\operatorname{Fix} \Phi$. To see that $\bar{X}$ is linearly independent, assume

$$\sum_{x \in X^c} \alpha_x \bar{x} = 0$$

for some coefficients $\alpha_x \in R$. Then one has $0 = \sum_{x \in X^c} \alpha_x \Phi^\infty x = \Phi^\infty \sum_{x \in X^c} \alpha_x x$. Since $\sum_{x \in X^c} \alpha_x x \in C(X^c)$, and $\Phi_{|C(X^c)} : C(X^c) \to \operatorname{Fix} \Phi$ is an isomorphism due to Proposition 8.3.1, we conclude that $\sum_{x \in X^c} \alpha_x x = 0$. This in turn implies $\alpha_x = 0$ for all $x \in X^c$. Therefore, the set $\bar{X}$ is indeed a basis for $\operatorname{Fix} \Phi$.

In order to see (ii), we observe that $\partial \bar{x} \in \operatorname{Fix} \Phi$, since Φ is a chain map. Thus, the representation in (8.12) is an immediate consequence of (i). Finally, a direct application of Proposition 3.5.3 shows that the pair $(\bar{X}, \bar{\kappa})$ is a Lefschetz complex, and clearly $C(\bar{X}) = \operatorname{Fix} \Phi$, which proves (iii). □

The Lefschetz complex $(\bar{X}, \bar{\kappa})$ will take the role of the Conley complex associated with the combinatorial gradient vector field $\mathcal{V}$. For this, and as a last step, it is necessary to identify a natural partial order on its cells. This can be accomplished as follows.

Recall that C denotes the collection of all critical cells of $\mathcal{V}$. By the definition of $\bar{X}$, the map $X^c \ni x \mapsto \bar{x} \in \bar{X}$ is a surjection. It follows from Proposition 8.3.1 that it is an injection, therefore a bijection. Hence, also the map $C \ni \{x\} \mapsto \bar{x} \in \bar{X}$ is a bijection, and this enables us to carry over the partial order $\leq_{\mathcal{V}}$ from $C \subset \mathcal{V}$ to the Lefschetz complex $\bar{X}$. Thus, for arbitrary elements $\bar{x}, \bar{y} \in \bar{X}$, we write $\bar{x} \leq_{\mathcal{V}} \bar{y}$ if one has $\{x\} \leq_{\mathcal{V}} \{y\}$. This gives the following result.

Proposition 8.3.3 *Assume that $\mathcal{V}$ is a combinatorial gradient vector field on a regular Lefschetz complex X. Then the above-defined partial order $\leq_{\mathcal{V}}$ in $\bar{X}$ is a natural partial order on the Lefschetz complex $(\bar{X}, \bar{\kappa})$.*

Proof We begin by proving that the order $\leq_{\mathcal{V}}$ in $\bar{X}$ is admissible, that is, the validity of $\bar{y} \leq_{\bar{\kappa}} \bar{x}$ for $x, y \in X^c$ implies $\bar{y} \leq_{\mathcal{V}} \bar{x}$. Since the partial order $\leq_{\bar{\kappa}}$ is the transitive closure of $\prec_{\bar{\kappa}}$, it clearly suffices to prove that $\bar{y} \prec_{\bar{\kappa}} \bar{x}$ implies $\bar{y} \leq_{\mathcal{V}} \bar{x}$. Thus, assume that $x, y \in X^c$ and $\bar{y} \prec_{\bar{\kappa}} \bar{x}$. Then, $\bar{\kappa}(\bar{x}, \bar{y}) \neq 0$. By Proposition 8.3.2, we have $\partial\bar{x} = \sum_{z \in X^c} a_{xz}\bar{z}$, with $a_{xz} = \bar{\kappa}(\bar{x}, \bar{z})$. In addition, Propositions 8.2.2 and 8.2.3 imply $\bar{z} = \Phi^{\infty} z = z + r_{z,\infty}$, together with the inclusion $|r_{z,\infty}| \subset X^+ \cap |\,[z]^{<_{\mathcal{V}}}|$. Since we also have $y \in X^c \subset X \setminus X^+$, one further obtains

$$
\begin{aligned}
\langle \partial\bar{x}, y\rangle = \sum_{z \in X^c} a_{xz}\langle \bar{z}, y\rangle &= \sum_{z \in X^c} a_{xz}\left(\langle z, y\rangle + \langle r_{z,\infty}, y\rangle\right) \\
&= \sum_{z \in X^c} a_{xz}\langle z, y\rangle = a_{xy} = \bar{\kappa}(\bar{x}, \bar{y}) \neq 0.
\end{aligned}
$$

It follows that

$$
y \in |\partial\bar{x}| = |\partial\Phi^{\infty} x| = |\Phi^{\infty}\partial x| = \left|\Phi^{\infty} \sum_{w \in |\partial x|} \kappa(x, w) w\right| \subset \bigcup_{w \in |\partial x|} |\Phi^{\infty} w|.
$$

Thus, $y \in |\Phi^{\infty} u|$ for some $u \in X$ such that $\kappa(x, u) \neq 0$. From Proposition 8.2.1, one now obtains $[y] \leq_{\mathcal{V}} [u]$. Since $\kappa(x, u) \neq 0$, we further have $u \in \operatorname{cl} x$, which in turn implies $u \in [u] \cap \operatorname{cl}[x]$. In consequence, $[u] \leq_{\mathcal{V}} [x]$ and $\{y\} = [y] \leq_{\mathcal{V}} [x] = \{x\}$. Thus, by the definition of $\leq_{\mathcal{V}}$ in $\bar{X}$, we get $\bar{y} \leq_{\mathcal{V}} \bar{x}$.

In order to see that $\leq_{\mathcal{V}}$ is indeed natural, we still need to verify the implication in (6.3). Thus, let $\bar{x}, \bar{y} \in \bar{X}$ satisfy both the inequality $\bar{x} \leq_{\mathcal{V}} \bar{y}$ and $\dim \bar{x} = \dim \bar{y}$. Then both $\{x\} \leq_{\mathcal{V}} \{y\}$ and $\dim x = \dim y$ hold as well. Yet this implies that one cannot have $\{x\} <_{\mathcal{V}} \{y\}$, because in this case, and in view of Proposition 8.1.2(ii), one deduces the strict inequality $\dim x = \dim\{x\} < \dim\{y\} = \dim y$. It follows that both $\{x\} = \{y\}$ and $\bar{x} = \bar{y}$ are satisfied, which proves that the order is natural.

□

Note that the partial order on $\bar{X}$ guaranteed by the proposition does not have to be the native partial order of the Lefschetz complex $(\bar{X}, \bar{\kappa})$. For this, consider a simplicial complex with three edges forming a triangle, and let the Forman gradient vector field $\mathcal{V}$ consist of one critical edge, an opposing critical vertex, and two Forman vectors. Then in the native order of $\bar{X}$, the critical vertex does not lie below the critical edge due to cancelation, while in the order induced by $\leq_{\mathcal{V}}$ it does.

In the next section of the book, we consider $\bar{X}$ as a poset ordered by the natural partial order $\leq_{\mathcal{V}}$ in $\mathcal{V}$, which has been transferred to $\bar{X}$ via the map $\{x\} \mapsto \bar{x}$. In addition, the triple $(\bar{X}, C(\bar{X}), \partial^{\bar{\kappa}})$ is considered as a natural filtration of $\bar{X}$.

8.4 Conley Complex and Unique Connection Matrix

Over the previous three sections, we have considered combinatorial gradient vector fields $\mathcal{V}$ on a regular Lefschetz complex X. Through the use of the associated combinatorial flow Φ, this study culminated in the creation of a new Lefschetz complex $\bar{X}$, together with its associated singleton partition. In fact, we have introduced $(\bar{X}, C(\bar{X}), \partial^{\bar{\kappa}})$ as a natural filtration of $\bar{X}$. We will now show that this filtration is the Conley complex associated with $\mathcal{V}$ and that it has a uniquely determined connection matrix. First, however, we need an auxiliary result, which allows us to recognize the combinatorial flow Φ and its iterates as filtered morphisms which are filtered chain homotopic to the identity.

Proposition 8.4.1 *Assume that $\mathcal{V}$ is a combinatorial gradient vector field on a regular Lefschetz complex X. Then we have a well-defined filtered morphism*

$$(\mathrm{id}_{\mathcal{V}}, \Phi) : (\mathcal{V}, C(X), \partial^{\kappa}) \to (\mathcal{V}, C(X), \partial^{\kappa}).$$

Moreover, $(\mathrm{id}_{\mathcal{V}}, \Phi)^n = (\mathrm{id}_{\mathcal{V}}, \Phi^n)$ is filtered chain homotopic to the identity morphism $\mathrm{id}_{(\mathcal{V}, C(X))}$ for every $n \in \mathbb{N}$. In particular, $(\mathrm{id}_{\mathcal{V}}, \Phi^{\infty})$ is filtered chain homotopic to $\mathrm{id}_{(\mathcal{V}, C(X))}$.

Proof The combinatorial flow Φ is a chain map, and $\mathrm{id}_{\mathcal{V}} : (\mathcal{V}, \mathcal{V}_{\star}) \to (\mathcal{V}, \mathcal{V}_{\star})$ is a morphism in DPSET. Hence, to prove that $(\mathrm{id}_{\mathcal{V}}, \Phi)$ is a well-defined filtered morphism, we only have to verify that the combinatorial flow Φ is $\mathrm{id}_{\mathcal{V}}$-filtered. We will do so in the following by establishing property (4.6) of Corollary 4.1.4. Consider the down set $\mathcal{L} \in \mathrm{Down}(\mathcal{V})$ and set $L := |\mathcal{L}|$. Then $C(X)_{\mathcal{L}} = \bigoplus_{V \in \mathcal{L}} C(V) = C(L)$, and we need to verify that $\Phi(C(L)) \subset C(L)$. But this follows from Proposition 8.1.8, because L is readily seen to be $\mathcal{V}$-compatible and closed by Proposition 6.1.2(i).

Since by Proposition 4.5.1 filtered chain homotopy between morphisms is preserved by composition, in order to prove that $(\mathrm{id}_{\mathcal{V}}, \Phi)^n = (\mathrm{id}_{\mathcal{V}}, \Phi^n)$ is filtered homotopic to the identity morphism, it suffices to prove that $(\mathrm{id}_{\mathcal{V}}, \Phi)$ is filtered homotopic to the identity morphism. By the definition of Φ, we have $\Phi - \mathrm{id}_{C(X)} = \partial\Gamma + \Gamma\partial$. Thus, one only needs to verify that Γ is $\mathrm{id}_{\mathcal{V}}$-filtered. Hence, assume that $\Gamma_{VW} \neq 0$ for some $V, W \in \mathcal{V}$. Then, there exists a $w \in W$ such that $\pi_V(\Gamma w) \neq 0$. In particular, one has $\Gamma w \neq 0$, which implies both $w = w^- \neq w^+$ and $\Gamma w = -\kappa(w^+, w^-)^{-1} w^+$. It follows that $w^+ \in V$. Thus $V = W = \mathrm{id}_{\mathcal{V}}(W)$, which proves that Γ is $\mathrm{id}_{\mathcal{V}}$-filtered, in fact, even graded. Finally, since the identity $\Phi^{\infty} = \Phi^n$ holds for large $n \in \mathbb{N}$, it follows that also $(\mathrm{id}_{\mathcal{V}}, \Phi^{\infty})$ is filtered chain homotopic to $\mathrm{id}_{(\mathcal{V}, C(X))}$. □

The next two results show that $(\bar{X}, C(\bar{X}), \partial^{\bar{\kappa}})$ is indeed a Conley complex of the combinatorial gradient vector field $\mathcal{V}$. First, we construct a filtered morphism from $(\mathcal{V}, C(X), \partial^{\kappa})$ to the Conley complex.

Proposition 8.4.2 *In the situation of the last proposition, we have a well-defined filtered morphism*

$$(\alpha, \varphi) : (\mathcal{V}, C(X), \partial^{\kappa}) \to (\bar{X}, C(\bar{X}), \partial^{\bar{\kappa}})$$

with $\alpha : \bar{X} \to \mathcal{V}$ *defined for cells* $x \in X^c$ *by* $\alpha(\bar{x}) := \{x\}$ *and* $\varphi : C(X) \to C(\bar{X})$ *defined for chains* $c \in C(X)$ *by* $\varphi(c) := \Phi^{\infty}(c)$.

Proof By Theorem 6.2.5(i), we have the equality $\bar{X}_{\star} = \bar{X}$. Thus, the poset filtered chain complex $(\bar{X}, C(\bar{X}), \partial^{\bar{\kappa}})$ is peeled. Furthermore, the map α is a morphism in DSET, and since we consider $\bar{X}$ as ordered by the natural order $\leq_{\mathcal{V}}$, the map α is trivially order preserving, hence also a morphism in DPSET. Obviously, the map φ is a chain map. We will now show that φ is α-filtered by verifying property (4.5) of Proposition 4.1.3. Let $\mathcal{L} \in \mathrm{Down}(\mathcal{V})$ and define $L := |\mathcal{L}|$. Then we have the identity $C(X)_{\mathcal{L}} = \bigoplus_{V \in \mathcal{L}} C(V) = C(L)$. Observe that $\alpha^{-1}(\mathcal{L})$ is a down set in $\bar{X}$. Indeed, if $\bar{x} \in \alpha^{-1}(\mathcal{L})$ and $\bar{y} \leq_{\mathcal{V}} \bar{x}$, then $\{y\} \leq_{\mathcal{V}} \{x\} = \alpha(\bar{x}) \in \mathcal{L}$, which implies the inclusions $\alpha(\bar{y}) = \{y\} \in \mathcal{L}$ and $\bar{y} \in \alpha^{-1}(\mathcal{L})$. Hence, $\alpha^{-1}(\mathcal{L})^{\leq} = \alpha^{-1}(\mathcal{L})$. It follows that $C(\bar{X})_{\alpha^{-1}(\mathcal{L})^{\leq}} = C(\bar{X})_{\alpha^{-1}(\mathcal{L})} = \bigoplus_{\{u\} \in \mathcal{L}} R\bar{u}$. Thus, in our situation, we have to verify condition (4.5) in the form

$$\Phi^{\infty}(C(L)) \subset \bigoplus_{\{u\} \in \mathcal{L}} R\bar{u}. \tag{8.13}$$

In order to establish (8.13), consider a chain $c \in C(L)$, and let $\bar{c} := \varphi(c) = \Phi^{\infty}(c)$. In view of Proposition 8.2.2, one has

$$\bar{c} = \sum_{u \in X^c} a_u \bar{u} = \sum_{u \in X^c} a_u u + \sum_{u \in X^c} a_u r_{u,\infty} \tag{8.14}$$

for some coefficients $a_u \in R$ and chains $r_{u,\infty}$ satisfying $|r_{u,\infty}| \cap X^c = \varnothing$, as well as the inclusion

$$|r_{u,\infty}| \subset |[u]^{<_{\mathcal{V}}}|. \tag{8.15}$$

Consider a critical cell $u \in X^c$ such that $\{u\} \notin \mathcal{L}$, i.e., one has $u \notin L$. Since by Proposition 8.1.8 we have $\varphi(C(L)) = \Phi^{\infty}(C(L)) \subset C(L)$, it then follows that the identity $\langle \bar{c}, u \rangle = 0$ holds. But now equation (8.14) gives $\langle \bar{c}, u \rangle = a_u$, because (8.15) implies $\langle r_{u,\infty}, u \rangle = 0$. Therefore, we have $a_u = 0$ for $u \notin L$, which yields

$$\bar{c} = \sum_{u \in L} a_u \bar{u} \in \bigoplus_{u \in L} R\bar{u}.$$

This verifies (8.13) and completes the proof. □

In our next result, we introduce a candidate for the inverse morphism in PFCC to the morphism (α, φ) from the last proposition.

Proposition 8.4.3 *In the situation of the last proposition, we have a well-defined filtered morphism*

$$(\beta, \psi) : (\bar{X}, C(\bar{X}), \partial^{\bar{\kappa}}) \to (\mathcal{V}, C(X), \partial^{\kappa})$$

with $\beta : \mathcal{V} \nrightarrow \bar{X}$ *defined for cells* $x \in X^c$ *by* $\beta(\{x\}) := \bar{x}$ *and* $\psi : C(\bar{X}) \to C(X)$ *defined for chains* $c \in C(X)$ *by* $\psi(c) := c$.

Proof As we already mentioned earlier, by Theorem 6.2.5(i), we have $\bar{X}_\star = \bar{X}$. Therefore, β is a morphism in DSET. Moreover, since we consider $\bar{X}$ as ordered by the natural order $\leq_{\mathcal{V}}$, the map β is trivially order preserving, hence also a morphism in DPSET. Obviously, the map ψ is a chain map. We will show that ψ is in fact β-filtered by checking again property (4.5) of Proposition 4.1.3. Consider a down set $A \in \operatorname{Down}(\bar{X})$. Clearly, we have $C(\bar{X})_A = C(A)$. If we define $\mathcal{L} := \beta^{-1}(A)$, then in general $\mathcal{L}$ is not a down set in $\mathcal{V}$, but $\mathcal{L}^{\leq_{\mathcal{V}}}$ is. We therefore consider $L := |\mathcal{L}^{\leq_{\mathcal{V}}}|$ and get $C(X)_{\beta^{-1}(A)^{\leq_{\mathcal{V}}}} = C(X)_{\mathcal{L}^{\leq_{\mathcal{V}}}} = C(L)$. Thus, in our case, condition (4.5) has to be verified in the form

$$\psi(C(A)) = C(A) \subset C(L). \tag{8.16}$$

Consider first a chain $\bar{x} \in A$ for a critical cell $x \in X^c$. Then $\beta(\{x\}) = \bar{x} \in A$, which means that $\{x\} \in \beta^{-1}(A) = \mathcal{L}$ and $|x| = \{x\} = [x] \in \mathcal{L} \subset \mathcal{L}^{\leq_{\mathcal{V}}}$. It follows from Proposition 8.2.2 that we have the representation $\bar{x} = \Phi^\infty x = x + r_{x,\infty}$, where the inclusion $|r_{x,\infty}| \subset X^+ \cap |\,[x]^{<_{\mathcal{V}}}\,|$ holds. Hence, one obtains that

$$|\bar{x}| \subset |x| \cup |r_{x,\infty}| \subset [x] \cup |[x]^{<_{\mathcal{V}}}| = |\,[x]^{\leq_{\mathcal{V}}}\,| \subset |\mathcal{L}^{\leq_{\mathcal{V}}}| = L$$

is satisfied, because $[x] \in \mathcal{L} \subset \mathcal{L}^{\leq_{\mathcal{V}}}$ and the set $\mathcal{L}^{\leq_{\mathcal{V}}}$ is a down set. This immediately shows that $\bar{x} \in C(L)$. Since every chain in $C(A)$ is a linear combination of chains $\bar{x}$ in A with $x \in X^c$, the inclusion (8.16) follows. □

The next result combines Propositions 8.4.2 and 8.4.3 to show that the filtered chain complexes $(\mathcal{V}, C(X), \partial^{\kappa})$ and $(\bar{X}, C(\bar{X}), \partial^{\bar{\kappa}})$ are indeed elementary filtered chain homotopic.

Theorem 8.4.4 (Existence of the Conley Complex) *Assume that* $\mathcal{V}$ *is a combinatorial gradient vector field on a regular Lefschetz complex* X*. Then the filtered morphisms* (α, φ) *and* (β, ψ) *defined in the last two propositions are mutually inverse elementary filtered chain equivalences. In particular, the poset filtered chain complexes* $(\mathcal{V}, C(X), \partial^{\kappa})$ *and* $(\bar{X}, C(\bar{X}), \partial^{\bar{\kappa}})$ *are elementary filtered chain homotopic, and the latter is a representation of the former, that is, its Conley complex.*

Proof Clearly, one has the identity $\beta\alpha = \mathrm{id}_{\bar{X}}$. Moreover, if $c \in \operatorname{Fix} \Phi$, then one obtains $\Phi^\infty(c) = c$. Therefore, the equality $\varphi\psi = \mathrm{id}_{\operatorname{Fix}\Phi} = \mathrm{id}_{C(\bar{X})}$ holds, and

$$(\alpha, \varphi) \circ (\beta, \psi) = (\beta\alpha, \varphi\psi) = (\mathrm{id}_{\bar{X}}, \mathrm{id}_{C(\bar{X})}) = \mathrm{id}_{(\bar{X}, C(\bar{X}))}\,.$$

In other words, $(\alpha, \varphi) \circ (\beta, \psi)$ is filtered homotopic to $\mathrm{id}_{(C, C(\bar{X}))}$ via the zero filtered homotopy. In the opposite direction, we have

$$(\beta, \psi) \circ (\alpha, \varphi) = (\alpha\beta, \psi\varphi) = (\mathrm{id}_{\mathcal{V}|C}, \Phi^{\infty}). \tag{8.17}$$

Since $\mathcal{V}_{\star} = C$, we see that $(\mathrm{id}_{\mathcal{V}|C}, \Phi^{\infty}) \sim_e (\mathrm{id}_{\mathcal{V}}, \Phi^{\infty})$. Hence, it follows from Proposition 8.4.1 and (8.17) that $(\alpha, \varphi) \circ (\beta, \psi)$ is filtered chain homotopic to the identity $\mathrm{id}_{(\mathcal{V}, C(X))}$, and also elementary filtered chain homotopic—and this completes the proof of the theorem. □

After these preparations, we have finally reached the main result of this chapter, which establishes the uniqueness of the connection matrix for gradient combinatorial vector fields on regular Lefschetz complexes.

Theorem 8.4.5 (Uniqueness of the Connection Matrix in the Gradient Case) *Let $\mathcal{V}$ be a combinatorial gradient vector field on a regular Lefschetz complex X. Then C, that is, the collection of critical vectors of $\mathcal{V}$, is a Morse decomposition of $\mathcal{V}$. It has exactly one connection matrix which coincides with the $(\bar{X}, \bar{X})$-matrix of $\partial^{\kappa}_{|\mathrm{Fix}\,\Phi}$, up to a graded similarity.*

Proof It follows from Proposition 8.1.1 that C is a Morse decomposition of $\mathcal{V}$ and that the partition induced on the complex X by this Morse decomposition is the vector field $\mathcal{V}$. Thus, by Definition 7.3.1, the Conley complex of C is the Conley complex of $(\mathcal{V}, C(X), \partial^{\kappa})$. From Corollary 5.2.7 and Theorem 8.4.4, we obtain that the Conley complex of $(\mathcal{V}, C(X), \partial^{\kappa})$ is isomorphic in PFCc to the Conley complex of $(\bar{X}, C(\bar{X}), \partial^{\bar{\kappa}})$, with $\bar{X}$ ordered by a natural partial order. Since, in view of Proposition 8.3.2 and Definition 6.2.4, the latter is the Conley complex of a natural filtration of a Lefschetz complex, the conclusion now follows directly from Theorem 6.2.5(ii). □

In addition, the above result allows us to easily detect the existence of connecting orbits in the setting of gradient combinatorial vector fields, that is, we generalize Theorem 7.3.6 to connection matrices with coefficients not necessarily in a field.

Theorem 8.4.6 (Existence of Connecting Orbits) *Let $\mathcal{V}$ be a gradient combinatorial vector field on a regular Lefschetz complex X. Furthermore, suppose that for a pair of critical cells $x, y \in X^c$, the associated connection matrix guaranteed by Theorem 8.4.5 has a nonzero entry in the $(\bar{x}, \bar{y})$-position. Then there exists a path with respect to the multivalued map $\Pi_{\mathcal{V}}$ from y to x, i.e., a connecting orbit.*

Proof In view of Theorem 8.4.5, the chain $\partial\bar{y}$ has to contain the basis element $\bar{x}$ in its representation with respect to the basis of $\mathrm{Fix}\,\Phi$ guaranteed by Proposition 8.3.2(i). Moreover, due to the representation of $\bar{x} = \Phi^{\infty}(x)$ given in Proposition 8.2.2, one has in fact the inclusion $x \in |\partial\bar{y}| = |\partial\Phi^{\infty}(y)|$. Using a simple inductive argument and Proposition 8.1.9(i) then immediately establishes the existence of a path from the cell y to the cell x. This completes the proof of the theorem. □

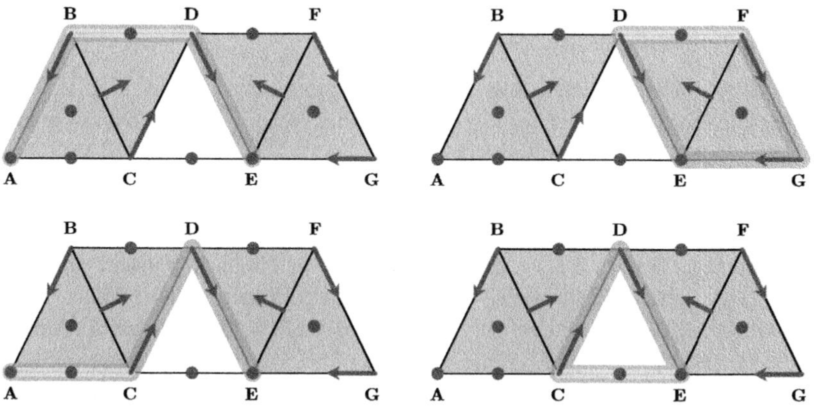

Fig. 8.4 *The stabilized combinatorial flow.* For the gradient combinatorial vector field $\mathcal{V}_3$ introduced in the third row of Fig. 2.6, the above four panels show the images $\Phi^{\infty}_{\mathcal{V}_3}(x)$ for the one-dimensional critical cells x. The chains $\Phi^{\infty}_{\mathcal{V}_3}(\mathbf{BD})$, $\Phi^{\infty}_{\mathcal{V}_3}(\mathbf{DF})$, $\Phi^{\infty}_{\mathcal{V}_3}(\mathbf{AC})$, and $\Phi^{\infty}_{\mathcal{V}_3}(\mathbf{CE})$ are shown in blue from top left to bottom right, respectively

We close this chapter with the following three examples. In the first of these, it will be demonstrated how the connection matrices in Table 2.2 can be determined using Theorem 8.4.5.

Example 8.4.7 (Three Forman Gradient Vector Fields $\lhd$ 8.2.4) Consider the gradient combinatorial Forman vector field $\mathcal{V}_3$ which was introduced in the third row of Fig. 2.6. By following the arguments in Example 8.2.4, one can easily show that $\Phi^{\infty}_{\mathcal{V}_3}(x) = x$ for all zero-dimensional critical cells x. Moreover, we have both $\Phi^{\infty}_{\mathcal{V}_3}(\mathbf{ABC}) = \mathbf{ABC} + \mathbf{BCD}$ and $\Phi^{\infty}_{\mathcal{V}_3}(\mathbf{EFG}) = \mathbf{EFG} + \mathbf{DEF}$. For the four one-dimensional critical cells y, their images $\Phi^{\infty}_{\mathcal{V}_3}(y)$ are illustrated in Fig. 8.4.

The images under the stabilized flow $\Phi^{\infty}_{\mathcal{V}_3}$ form the basis with respect to which one can determine the connection matrix. As before, we abbreviate these by $\bar{c} = \Phi^{\infty}_{\mathcal{V}_3}(c)$ for all critical cells $c \in X^c$. Using this notation, for example, in order to determine the last column of the third connection matrix in Table 2.2, we have to express the boundary $\partial\overline{\mathbf{EFG}}$ in terms of the chains $\bar{y}$ for the one-dimensional critical cells y, which leads to the identity

$$\partial\overline{\mathbf{EFG}} = \mathbf{DF} + \mathbf{FG} + \mathbf{EG} + \mathbf{DE} = \overline{\mathbf{DF}},$$

and this accounts for the single entry of 1 in the corresponding column given in Table 2.2. Similarly, since we have

$$\partial\overline{\mathbf{ABC}} = \mathbf{AB} + \mathbf{BD} + \mathbf{CD} + \mathbf{AC} = \overline{\mathbf{AC}} + \overline{\mathbf{BD}},$$

the edge **DE** which is contained in both of the last two chains cancels upon addition with respect to $\mathbb{Z}_2$-coefficients. This leads to the two nonzero entries in the second-to-last column of the last connection matrix in Table 2.2. The remaining columns can be determined analogously. ◊

Next, we briefly sketch how the above example can be used to determine connection matrices in the case of a combinatorial vector field which is not gradient. For this, we return to the setting of Fig. 2.2.

Example 8.4.8 (A Forman Vector Field with Periodic Orbit, ◁ 2.6.4) Consider again the combinatorial vector field $\mathcal{V}_0$ from Fig. 2.2, which has already been discussed in Examples 2.2.3 and 2.6.4. This vector field has a periodic orbit which traverses the vertices **C**, **D**, and **E**—and by breaking this periodic orbit into the union of two critical cells of dimensions 0 and 1, as well as the associated connecting orbits, we arrived at the three gradient systems $\mathcal{V}_k$ for $k = 1, 2, 3$ shown in Fig. 2.6.

We now claim that each of the connection matrices from Table 2.2 is in fact a connection matrix for $\mathcal{V}_0$. Note that this vector field has a Morse decomposition $\mathcal{M}_0$ in which one Morse set, say M_p, consists of the periodic orbit. Since the associated Conley index has the Conley polynomial $t + 1$, we now further subdivide this Morse set into a critical cell $M_{\tilde{p}_1}$ of index 1, one critical cell $M_{\tilde{p}_0}$ of index 0, as well as two Forman vectors, as shown in Fig. 2.6. This results in one of the combinatorial vector fields $\mathcal{V}_k$, which in turn leads to a refinement of the associated acyclic partitions used in Definition 7.3.1. But then a direct application of Propositions 6.3.5 and 6.3.7 shows that the Conley complex associated with $\mathcal{V}_k$ is also a Conley complex for $\mathcal{V}_0$. (Note in particular that the above introduction of the critical cells $M_{\tilde{p}_1}$ and $M_{\tilde{p}_0}$ implies that the Conley complex has a trivial boundary operator on the coarsened Morse set which includes the periodic orbit in $\mathcal{M}_0$.) In other words, the connection matrices shown in Table 2.2 are also connection matrices for $\mathcal{V}_0$.

In fact, even more is true. Not only are all three connection matrices from Table 2.2 connection matrices for $\mathcal{V}_0$, they are also pairwise nonequivalent in the sense of Definition 5.5.5, that is, the vector field $\mathcal{V}_0$ does not have a unique connection matrix. In the remainder of this example, we briefly sketch the necessary proof of nonequivalence.

For this, we only consider the vector fields $\mathcal{V}_1$ and $\mathcal{V}_2$. In view of Propositions 8.4.2 and 8.4.3, the transfer morphism $T : C' \to C''$ from the Conley complex $(\mathcal{M}_0, C', d')$ associated with $\mathcal{V}_1$ to the Conley complex $(\mathcal{M}_0, C'', d'')$ associated with $\mathcal{V}_2$ is defined via the formula

$$T(\bar{c}) = \Phi^{\infty}_{\mathcal{V}_2}(c) \quad \text{for} \quad \bar{c} = \Phi^{\infty}_{\mathcal{V}_1}(c) .$$

Furthermore, the matrix representations of the boundary operators d' and d'' can be found in the first two entries of Table 2.2.

As before we suppose that there exists a graded morphism f and a degree $+1$ homomorphism $G : C' \to C''$ with

$$T - f = d''G + Gd' .$$

We now evaluate both sides of this equation at the specific chain $\overline{\mathbf{DF}}$. The first matrix in Table 2.2 shows that $d'(\overline{\mathbf{DF}}) = 0$. Furthermore, quick glances at Figs. 8.2 and 8.3 imply that $T(\overline{\mathbf{DF}}) = \overline{\mathbf{DF}} + \overline{\mathbf{DE}}$. Finally, since f is a graded morphism, we have to have $f(\overline{\mathbf{DF}}) = \alpha\overline{\mathbf{DF}}$ for some $\alpha \in \mathbb{Z}_2$. Substituting these into the above identity relating T and f, one then obtains

$$\overline{\mathbf{DF}} + \overline{\mathbf{DE}} - \alpha\overline{\mathbf{DF}} = d''(G(\overline{\mathbf{DF}})) .$$

This equation, however, is impossible for $\alpha = 1$ due to the matrix form of d'' given in Table 2.2, since $(0, 0, 1, 0, 0, 0)^t$ (which corresponds to the chain $\overline{\mathbf{DE}}$) is not contained in the column space of d''. If instead we assume $\alpha = 0$, then the form of the matrix d'' implies the equality $G(\overline{\mathbf{DF}}) = \overline{\mathbf{EFG}}$, and since G is filtered, this further gives $\overline{\mathbf{EFG}} \leq \overline{\mathbf{DF}}$, a contradiction. In other words, the connection matrices arising from the vector fields $\mathcal{V}_1$ and $\mathcal{V}_2$ are not equivalent in the sense of Definition 5.5.5.

The nonequivalence proofs for $\mathcal{V}_1 \not\equiv \mathcal{V}_3$, as well as $\mathcal{V}_2 \not\equiv \mathcal{V}_3$, are similar and thus left to the reader. We would like to point out, however, that the first of these two cases requires a slightly more involved argument. ◇

As our last example, we return to the discussions begun in Examples 5.2.8 and 5.5.7, which center on the combinatorial flows in Fig. 7.1.

Example 8.4.9 (Nonuniqueness via Subdivision, ◁ 7.3.2) Consider finally the partition $\mathcal{E}_1 := \mathcal{E}_{\mathcal{M}_1}$ associated with the Morse decomposition $\mathcal{M}_1$ of the combinatorial vector field $\mathcal{V}_1$ in the middle of Fig. 7.1. In this simple case, it coincides with the vector field $\mathcal{V}_1$. Hence, the Hasse diagram of the partial order $\leq_{\mathcal{E}_1}$ is given by diagram (7.2). It follows that $\mathcal{V}_1$ is a gradient vector field, and by Theorem 8.4.5, the Morse decomposition $\mathcal{M}_1$ has exactly one connection matrix. One can verify that the morphisms (ε, h') and (ε^{-1}, g') constructed as in Example 5.2.8 are also filtered with respect to the filtration $(\mathcal{E}_1, C(X), \partial^\kappa)$. Therefore, arguing as in Example 5.2.8, we conclude that the Conley complex of $\mathcal{M}_1$ is the filtered chain complex described in Example 4.3.3 with the associated connection matrix (5.2). In fact, the homomorphism h' coincides with the homomorphism $\Phi_{\mathcal{V}_1}$, and we have $\Phi_{\mathcal{V}_1} = \Phi^\infty_{\mathcal{V}_1}$ in this case. An analogous argument shows that the Conley complex of $\mathcal{M}_2$ is the filtered chain complex in Example 5.5.7 with the associated connection matrix (5.29). Both Conley complexes are visualized in Fig. 8.5 as κ-subcomplexes of suitable simplicial complexes. ◇

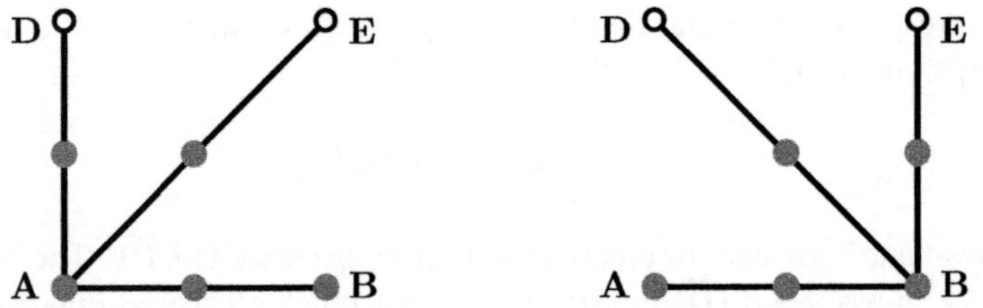

Fig. 8.5 Conley complexes $(\bar{X}, \partial^{\bar{\kappa}})$ for the Morse decompositions $\mathcal{M}_1$ (left) and $\mathcal{M}_2$ (right) of the combinatorial gradient vector fields $\mathcal{V}_1$ and $\mathcal{V}_2$ in Fig. 7.1, visualized as κ-subcomplexes of simplicial complexes

8.5 Taking Stock and the Next Steps

The main motivation for the present book was to carry over the classical theory of connection matrices to the case of combinatorial multivector fields on Lefschetz complexes. Yet, in the course of this, we achieved significantly more. By allowing for the change of poset in the underlying Morse decomposition, we now have a mechanism at hand to compare and ultimately classify connection matrices. In contrast to existing results, this makes it possible to give a precise meaning to the phenomenon of multiple connection matrices. As a welcome side effect, we could also shorten the connection matrix pipeline described in the introductory chapter. Combined with the recent development of concrete algorithms for the computation of connection matrices, as described in Sect. 2.7, this should considerably broaden their applicability in concrete examples.

Nevertheless, the results presented in this book are just a first step. There are a number of important open questions that have to be addressed in the future and that will be briefly outlined in the following:

- While our approach gives a precise definition for when two connection matrices are the same, verifying this equivalence is not immediately straightforward. For this, it is necessary to recognize a morphism as essentially graded. Are there easy ways to do this?
- Related to the previous point, are there easy ways to see that two connection matrices are really different, i.e., that one is in the situation of a combinatorial multivector field with multiple connection matrices? Are there easily computable invariants that can be used to detect this?
- What is the deeper meaning of the occurrence of multiple connection matrices? In the classical case, this is always seen as an indication of dynamical objects such as saddle-saddle connections. Is the same true in the situation of combinatorial multivector fields?
- In the classical case, transition matrices encode global bifurcations that can be detected using a change in connection matrices. Can this notion be extended to the combinatorial multivector case? In fact, we suppose that this is strongly related to our notion of transfer morphisms. In what way precisely? In order

to address this question, one will also have to establish a precise notion of continuation of connection matrices.

- Combinatorial dynamical systems induced by multivector fields constitute an analog of classical flows. Conley theory, originally developed for flows, was later extended to dynamical systems with discrete time, that is, dynamical systems obtained by iterating maps [34, 43] or multivalued maps [6] on Hausdorff topological spaces. The construction of connection matrices by Robbin & Salamon [44] applies to dynamical systems with continuous as well as discrete time under the assumption that the topological space is Hausdorff. Recently, Barmak, Mrozek, and Wanner [4] presented Conley theory for multivalued maps on finite topological spaces. Thus, it is natural to wonder whether connection matrix theory can be extended to this setting as well.
- Can every connection matrix be computed using the algorithms mentioned in Sect. 2.7? We already indicated in the previous section in Example 8.4.8 that in certain cases it is possible to break up recurrent Morse sets in such a way that the resulting combinatorial multivector field is actually gradient. Are the connection matrices obtained in this way all possible connection matrices? If not, is it possible to obtain the remaining ones algorithmically?
- The persistence-based connection matrix algorithm mentioned in Sect. 2.7 indicates the possibility of the interpretation of connection matrix theory in the language of persistence. Such a possibility might lead to applications of connection matrices which go beyond dynamics.
- The proof of the existence of the Conley complex (Theorem 5.3.2) and, implicitly, the existence of connection matrices requires field coefficients. However, as we emphasize at the beginning of this chapter, in the setting of Forman gradient vector fields, this assumption is not needed, because a Conley complex may be established via the subcomplex of chains fixed by the combinatorial flow Φ on chains (see Proposition 8.3.2 and Theorem 8.4.4). Forman's construction of the combinatorial flow Φ uses the fact that the only regular vectors are doubletons. A regular vector, that is, a doubleton multivector, considered as a Lefschetz complex has an associated chain complex which is chain homotopic to the zero complex. This is then used to construct the chain homotopy needed to prove Theorem 8.4.4. However, by the Acyclic Carrier Theorem (see [39, Theorems 13.3 and 13.4]), every regular multivector has its chain complex chain homotopic to the zero complex. Thus, it would be interesting to investigate to what extent Forman's construction of Φ and, in consequence, the construction of the Conley complex and the connection matrix may be generalized to gradient multivector fields.

Progress on any of these topics would only serve to extend the range of potential applications of this theory.

References

1. P. Alexandrov, Diskrete Räume. Mathematiceskii Sbornik (N.S.) **2**, 501–518 (1937)
2. J.A. Barmak, *Algebraic Topology of Finite Topological Spaces and Applications*, vol. 2032 of *Lecture Notes in Mathematics* (Springer, Berlin, Heidelberg, 2011)
3. J.A. Barmak, M. Mrozek, T. Wanner, A Lefschetz fixed point theorem for multivalued maps of finite spaces. Mathematische Zeitschrift **294**(3–4), 1477–1497 (2020)
4. J.A. Barmak, M. Mrozek, T. Wanner, Conley index for multivalued maps on finite topological spaces. Found. Comput. Math. (2024). Accepted for publication. arXiv:2310.03099v2
5. B. Batko, T. Kaczynski, M. Mrozek, T. Wanner, Linking combinatorial and classical dynamics: Conley index and Morse decompositions. Found. Comput. Math. **20**(5), 967–1012 (2020)
6. B. Batko, M. Mrozek, Weak index pairs and the Conley index for discrete multivalued dynamical systems. SIAM J. Appl. Dyn. Syst. **15**(2), 1143–1162 (2016)
7. J. Bezanson, A. Edelman, S. Karpinski, V.B. Shah, Julia: A fresh approach to numerical computing. SIAM Rev. **59**(1), 65–98 (2017)
8. G. Birkhoff, Rings of sets. Duke Math. J. **3**(3), 443–454 (1937)
9. E. Boczko, W.D. Kalies, K. Mischaikow, Polygonal approximation of flows. Topol. Appl. **154**(13), 2501–2520 (2007)
10. S.S. Cairns, A simple triangulation method for smooth manifolds. Bull. Amer. Math. Soc. **67**, 389–390 (1961)
11. C. Conley, *Isolated Invariant Sets and the Morse Index*, vol. 38 of *CBMS Regional Conference Series in Mathematics* (American Mathematical Society, Providence, RI, 1978)
12. T.K. Dey, M. Juda, T. Kapela, J. Kubica, M. Lipinski, M. Mrozek, Persistent homology of morse decompositions in combinatorial dynamics. SIAM J. Appl. Dyn. Syst. **18**, 510–530 (2019)
13. T.K. Dey, M. Lipiński, M. Mrozek, R. Slechta, Computing connection matrices via persistence-like reductions. SIAM J. Appl. Dyn. Syst. **23**(1), 81–97 (2024)
14. P. Dlotko, T. Kaczynski, M. Mrozek, T. Wanner, Coreduction homology algorithm for regular CW-complexes. Discrete Comput. Geometry **46**(2), 361–388 (2011)
15. H. Edelsbrunner, J.L. Harer, *Computational Topology*. (American Mathematical Society, Providence, 2010)
16. M. Eidenschink, *Exploring global dynamics: A numerical algorithm based on the Conley Index theory*. PhD thesis, Georgia Institute of Technology, 1995
17. R. Engelking, *General Topology* (Heldermann Verlag, Berlin, 1989)
18. R. Forman, Combinatorial vector fields and dynamical systems. Math. Z. **228**(4), 629–681 (1998)

M. Mrozek, T. Wanner, *Connection Matrices in Combinatorial Topological Dynamics*, SpringerBriefs in Mathematics,
https://doi.org/10.1007/978-3-031-87600-4

19. R. Forman, Morse theory for cell complexes. Adv. Math. **134**(1), 90–145 (1998)
20. R.D. Franzosa, The connection matrix theory for Morse decompositions. Trans. Amer. Math. Soc. **311**(2), 561–592 (1989)
21. S. Harker, K. Mischaikow, K. Spendlove, A computational framework for connection matrix theory. J. Appl. Comput. Topol. **5**(3), 459–529 (2021)
22. M. Jöllenbeck, V. Welker, Minimal resolutions via algebraic discrete Morse theory. Memoirs Amer. Math. Soc. **197**(923), vi+74 (2009)
23. T. Kaczynski, K. Mischaikow, M. Mrozek, *Computational Homology*, vol. 157 of *Applied Mathematical Sciences* (Springer, New York, 2004)
24. T. Kaczynski, M. Mrozek, M. Ślusarek, Homology computation by reduction of chain complexes. Comput. Math. Appl. **35**, 59–70 (1998)
25. T. Kaczynski, M. Mrozek, T. Wanner, Towards a formal tie between combinatorial and classical vector field dynamics. J. Comput. Dyn. **3**(1), 17–50 (2016)
26. K.P. Knudson, *Morse Theory: Smooth and Discrete* (World Scientific, 2015)
27. D. Kozlov, *Combinatorial Algebraic Topology*, vol. 21 of *Algorithms and Computation in Mathematics* (Springer, Berlin, 2008)
28. S. Lang, *Algebra*, vol. 211 of *Graduate Texts in Mathematics*, 3rd edn. (Springer, New York, 2002)
29. S. Lefschetz, *Algebraic Topology*. American Mathematical Society Colloquium Publications, v. 27 (American Mathematical Society, New York, 1942)
30. M. Lipinski, J. Kubica, M. Mrozek, T. Wanner, Conley-Morse-Forman theory for generalized combinatorial multivector fields on finite topological spaces. J. Appl. Comput. Topol. **7**(2), 139–184 (2023)
31. S. Maier-Paape, K. Mischaikow, T. Wanner, Structure of the attractor of the Cahn-Hilliard equation on a square. Int. J. Bifurcat. Chaos **17**(4), 1221–1263 (2007)
32. W.S. Massey, *A Basic Course in Algebraic Topology*. *Graduate Texts in Mathematics*, v. 127. (Springer, New York, 1991)
33. M.C. McCord, Singular homology groups and homotopy groups of finite topological spaces. Duke Math. J. **33**, 465–474 (1966)
34. M. Mrozek, Leray functor and cohomological Conley index for discrete dynamical systems. Trans. Amer. Math. Soc. **318**(1), 149–178 (1990)
35. M. Mrozek, Conley-Morse-Forman theory for combinatorial multivector fields on Lefschetz complexes. Found. Comput. Math. **17**(6), 1585–1633 (2017)
36. M. Mrozek, B. Batko, Coreduction homology algorithm. Discrete Comput. Geom. **41**(1), 96–118 (2009)
37. M. Mrozek, R. Srzednicki, J. Thorpe, T. Wanner, Combinatorial vs. classical dynamics: Recurrence. Commun. Nonlinear Sci. Numer. Simul. **108**, Paper No. 106226, 30 pp. (2022)
38. M. Mrozek, T. Wanner, Creating semiflows on simplicial complexes from combinatorial vector fields. J. Differ. Equ. **304**, 375–434 (2021)
39. J.R. Munkres, *Elements of Algebraic Topology* (Addison-Wesley, Boston, MA, 1984)
40. J.R. Munkres, *Topology* (Prentice Hall, Upper Saddle River, NJ, 2000)
41. J.F. Reineck, The connection matrix in Morse-Smale flows. Trans. Amer. Math. Soc. **322**(2), 523–545 (1990)
42. J.F. Reineck, The connection matrix in Morse-Smale flows. II. Trans. Amer. Math. Soc. **347**(6), 2097–2110 (1995)
43. J.W. Robbin, D. Salamon, Dynamical systems, shape theory and the Conley index. Ergodic Theory Dynam. Syst. **8**, 375–393 (1988)
44. J.W. Robbin, D.A. Salamon, Lyapunov maps, simplicial complexes and the Stone functor. Ergodic Theory Dynam. Syst. **12**(1), 153–183 (1992)
45. E. Sköldberg, Morse theory from an algebraic viewpoint. Trans. Amer. Math. Soc. **358**, 115–129 (2005)
46. K. Spendlove, *Computational Connection Matrix Theory*. PhD thesis, Rutgers University, 2019
47. T. Stephens, T. Wanner, Rigorous validation of isolating blocks for flows and their Conley indices. SIAM J. Appl. Dyn. Syst. **13**(4), 1847–1878 (2014)

48. J. Thorpe, T. Wanner, Global dynamics via multivector fields on Lefschetz complexes. In preparation (2025)
49. T. Wanner, ConleyDynamics.jl: A Julia package for multivector dynamics on Lefschetz complexes (2024). https://github.com/almost6heads/ConleyDynamics.jl
50. D. Woukeng, D. Sadowski, J. Leśkiewicz, M. Lipiński, T. Kapela, Rigorous computation in dynamics based on topological methods for multivector fields. J. Appl. Comput. Topol. **8**(4), 875–908 (2024)

Index

M. Mrozek, T. Wanner, *Connection Matrices in Combinatorial Topological Dynamics*, SpringerBriefs in Mathematics,
https://doi.org/10.1007/978-3-031-87600-4

GPSR Compliance

The European Union's (EU) General Product Safety Regulation (GPSR) is a set of rules that requires consumer products to be safe and our obligations to ensure this.

If you have any concerns about our products, you can contact us on ProductSafety@springernature.com

In case Publisher is established outside the EU, the EU authorized representative is:

Springer Nature Customer Service Center GmbH
Europaplatz 3
69115 Heidelberg, Germany

Batch number: 08693348

Printed by Printforce, the Netherlands